Decision Engineering

Series Editor

Professor Rajkumar Roy
Department of Enterprise Integration
School of Industrial and Manufacturing Science
Cranfield University
Cranfield
Bedford
MK43 0AL
UK

Other titles published in this series

Cost Engineering in Practice
John McIlwraith

IPA – Concepts and Applications in Engineering
Jerzy Pokojski

Strategic Decision Making
Navneet Bhushan and Kanwal Rai

Product Lifecycle Management
John Stark

From Product Description to Cost: A Practical Approach
Volume 1: The Parametric Approach
Pierre Foussier

From Product Description to Cost: A Practical Approach
Volume 2: Building a Specific Model
Pierre Foussier

Decision-Making in Engineering Design
Yotaro Hatamura

Composite Systems Decisions
Mark Sh. Levin

Intelligent Decision-making Support Systems
Jatinder N.D. Gupta, Guisseppi A. Forgionne and Manuel Mora T.

Knowledge Acquisition in Practice
N.R. Milton

Global Product: Strategy, Product Lifecycle Management and the Billion Customer Question
John Stark

DRILL HALL LIBRARY
MEDWAY

Andrew Greasley

Enabling a Simulation Capability in the Organisation

Andrew Greasley, MBA, PhD, FHEA
Technology and Operations Management
Aston Business School
Aston University
Birmingham
B4 7ET
UK

ISBN 978-1-84800-168-8   ISBN 978-1-84800-169-5 (eBook)

DOI 10.1007/978-1-84800-169-5

Decision Engineering Series ISSN 1619-5736

British Library Cataloguing in Publication Data
Greasley, Andrew
  Enabling a simulation capability in the organisation. -
  (Decision engineering)
  1. Decision making - Simulation methods 2. Operations
  research
  I. Title
  658.4'0352
ISBN-13: 9781848001688

Library of Congress Control Number: 2008923782

© 2008 Springer-Verlag London Limited

Apart from any fair dealing for the purposes of research or private study, or criticism or review, as permitted under the Copyright, Designs and Patents Act 1988, this publication may only be reproduced, stored or transmitted, in any form or by any means, with the prior permission in writing of the publishers, or in the case of reprographic reproduction in accordance with the terms of licences issued by the Copyright Licensing Agency. Enquiries concerning reproduction outside those terms should be sent to the publishers.

The use of registered names, trademarks, etc. in this publication does not imply, even in the absence of a specific statement, that such names are exempt from the relevant laws and regulations and therefore free for general use.

The publisher makes no representation, express or implied, with regard to the accuracy of the information contained in this book and cannot accept any legal responsibility or liability for any errors or omissions that may be made.

Cover design: eStudio Calamar S.L., Girona, Spain

Printed on acid-free paper

9 8 7 6 5 4 3 2 1

springer.com

Dedicated to
Philly, TJ,
Hugo and Humphrey

Preface

The aim of this book is to help enable the use of the technique of simulation modelling in the organisation. There is an emphasis in the book on the implementation of the technique in organisations, rather than a detailed treatment of the mechanics of simulation software execution or the statistical analysis undertaken during a study. The focus in this text is on recent changes in the way simulation is used which have led to its wider use and provide the potential for a continued growth in use.

Chapter 1 provides an overview of the technique and gives details of where the technique can be applied in the organisation.

Chapter 2 provides evidence of previous and current survey research showing the use of simulation and the challenges ahead if usage is to be increased. A particular area of growth in the use of simulation is in service applications. This has led to an interest in the modelling of human behaviour in addition to the traditional simulation of material and information processes. A framework is presented indicating approaches to the challenge of modelling people in organisations to provide guidance for simulation practitioners.

Chapter 3 examines the physical and human resources that are necessary to enable a simulation capability within the organisation.

Chapter 4 outlines the steps required in undertaking a simulation project. In order to use simulation successfully a structured process must be followed. This chapter aims to show that simulation is about more than just the purchase and use of a software package but a range of skills are required by the simulation team. These include project management, client liaison, statistical skills, modelling skills and the ability to understand and map out organisational processes.

Chapters 5, 6 and 7 provide extensive case study research of the use of the simulation technique in the organisation. The material is organised under three themes that emerged from the case study investigations.

Chapter 5 covers the use of simulation within a process-centred change methodology. Business Process Simulation (BPS) is entering the mainstream of process improvement tools, in part on the back of process-centred change methodologies such as Business Process Management. It is generally accepted that the process perspective can deliver benefits and BPS can improve the chance of success by providing a tool for analysis. The process-based change methodology can provide context to the simulation technique in that it connects the aims of the

BPS study to the strategic objectives of the organisation and incorporates the consideration of human factors in order to achieve successful implementation of redesigned processes. Conversely the ability of BPS to incorporate system variability, scenario analysis and visual display of process performance makes it a useful technique to provide a realistic assessment of the need for and results of change.

Chapter 6 covers the use of qualitative outcomes of a simulation intervention. Simulation is found to have the ability to facilitate knowledge through the day-to-day process of undertaking the study, for example collecting the data and mapping the processes, and providing qualitative outcomes, for example an animation of the system incorporating individual elements such as people and materials.

Chapter 7 covers the use of simulation in combination with other improvement techniques. The Activity-Based Costing approach allows the actual costs to be traced to activities and so enables better resource allocation decisions. Reversing the flow of information allows the user to assess the effect of a change in the activity level on costs. Simulation is also shown in use in conjunction with the technique of system dynamics. It is shown that the system dynamics approach is particularly appropriate in analysing factors impacting on the organisational context of a simulation study and thus could be used to maximise the benefits of simulation. Finally the technique of Data Envelopment Analysis is shown to be a useful addition to the toolkit of a simulation analyst in that it is able to rank the relative performance of units across multiple input and output measures.

Andrew Greasley, MBA PhD FHEA
November 2007

Acknowledgements

The author wishes to thank Anthony Doyle of Springer-Verlag London for his support of this text and the author also wishes to thank all the people involved in the preparation of this text for publication. The author is grateful for permission to reproduce material from previously published journal articles which appears in Chapters 5, 6 and 7 of this text.

Emerald Publishing has granted permission to publish the following articles in this text:

Greasley, A. (2004) A redesign of a road traffic accident reporting system using business process simulation, *Business Process Management Journal,* Vol. 10, No. 6, pp. 636–644. (Case Study 1)
Greasley, A. (2003) Using Business Process Simulation within a Business Process Reengineering approach, *Business Process Management Journal*, Vol 9 No. 4, pp. 408–420. (Case Study 2)
Greasley, A. (2004) Process Improvement within a HR Division at a UK Police Force, *International Journal of Operations and Production Management*, 24(2/3), pp. 230–240. (Case Study 3)
Greasley, A. (2004) The case for the organisational use of simulation, *Journal of Manufacturing Technology Management*, Vol. 15 No. 7, pp. 560–566. (Case Study 5)
Greasley, A. (2005) "Using system dynamics in a discrete-event simulation study of a manufacturing plant", *International Journal of Operations and Production Management,* Vol. 25 No. 6, pp. 534–548. (Case Study 8)

John Wiley and Sons Limited have granted permission to publish the following article in this text:

Greasley, A. (2000) Using Simulation to Assess the Service Reliability of a Train Maintenance Depot, *Quality and Reliability Engineering International*, 16(3), pp. 221–228. (Case Study 4)

The Operational Research Society have granted permission to publish the following articles in this text:

Greasley, A. (2000) A simulation analysis of arrest costs, *Journal of the Operational Research Society*, 51, pp. 162–167. (Case Study 7)

Greasley, A. (2005) Using DEA and simulation in guiding operating units to improved performance, *Journal of the Operational Research Society,* Vol. 56, pp. 727–731. (Case Study 9)

Contents

1 Introduction to Simulation 1
   Introduction 1
   What is Simulation Modelling? 1
   Simulation and Variability 2
      Variability 2
      Interdependence 3
   Why Use Simulation? 4
   Disadvantages of the Simulation Method 5
   Summary 6
   References 6

2 The Usage of Simulation 7
   Introduction 7
   Published Surveys of the Use of Simulation 7
   A Survey of Simulation Use in the UK 9
   Where in the Organisation is Simulation Used? 12
      Manufacturing Applications 12
      Service Applications 13
   Modelling Human Behaviour with Simulation 13
      Methods of Modelling Human Behaviour 14
      Choosing a Method to Model Human Behaviour 17
   Summary 17
   References 18

3 Acquiring the Resources for Simulation 21
   Introduction 21
   Steps in Introducing Simulation 21
      1. Select Simulation Sponsor 21
      2. Evaluate Potential Benefits of Simulation 22
      3. Estimate Resource Requirements 22
      4. Selecting the Simulation Software Type 23
      5. Selecting the Simulation Software Package 26

	6. Computer Hardware Requirements	27
	7. Training	27
	Summary	29
	References	29
4	**Steps in Building a Simulation Model**	**31**
	Introduction	31
	1. Formulate the Simulation Project Proposal	31
	Determining the Level of Usage of the Simulation Model	31
	Managing the Simulation Project	33
	The Simulation Project Proposal	36
	2. Data Collection	39
	Logic Data Required for the Process Map	39
	Additional Data Required for the Simulation Model	39
	3. Process Mapping	41
	Activity Cycle Diagrams	42
	Process Maps	42
	4. Modelling Input Data	42
	Less than 20 Data Points: Estimation	43
	20+ Data Points: Deriving a Theoretical Distribution	43
	200+ Data Points: Constructing an Empirical Distribution	44
	Historical Data Points	44
	5. Building the Model	44
	6. Validation and Verification	44
	Verification	45
	Validation	46
	7. Experimentation and Analysis	48
	Statistical Analysis for Terminating Systems	49
	Statistical Analysis for Non-terminating Systems	52
	8. Presentation of Results	54
	9. Implementation	55
	Organisational Context of Implementation	55
	Summary	57
	References	57
5	**Enabling Simulation – Simulation and Process Improvement Methodology**	**59**
	Introduction	59
	Case Study 1: A Redesign of a Road Traffic Accident Reporting System Using Business Process Simulation	61
	Introduction	61
	The Road Traffic Accident Case Study	61
	The Road Traffic Accident Business Process Simulation	63
	Discussion	66

Case Study 2: Using Business Process Simulation
Within a Business Process Reengineering Approach 67
    Introduction 67
    The Custody of Prisoner Process Case Study 67
    Discussion 72
Case Study 3: Process Improvement Within a HR Division
at a UK Police Force 75
    Introduction 75
    The HR Division Case Study 75
    Discussion 81
Summary 83
References 85

6   Enabling Simulation – Qualitative Simulation 89
Introduction 89
Case Study 4: Using Simulation Modelling to Assess Service
Reliability 89
    Introduction 89
    Case Study 90
    The Simulation Study 91
    Discussion 97
Case Study 5: The Case for the Organisational Use of Simulation 97
    Introduction 97
    Case Study 98
    The Simulation Study 99
    Simulation Model Analysis 100
    Simulation Study Results 103
    Discussion 103
Case Study 6: Using Simulation for Facility Design 106
    Introduction 106
    The Case Study 106
    The Simulation Study 110
    Discussion 114
Summary 115
References 116

7   Enabling Simulation – Simulation and OR Techniques 119
Introduction 119
Case Study 7: A Simulation Analysis of Arrest Costs 121
    Introduction 121
    Activity Based Costing – Committed and Flexible Resource 122
    The Simulation Study 123
    Discussion 129

Case Study 8: Using System Dynamics
in a Discrete-event Simulation Study 130
    Introduction 130
    The Manufacturing Process... 131
    The Discrete-event Simulation Study... 133
    The System Dynamics Study.. 136
    Discussion 140
Case Study 9: The Use of Data Envelopment Analysis
in a Discrete-event Simulation Study 141
    Introduction 141
    Preliminary Data Analysis.. 141
    The Initial DEA Assessment 142
    Further Investigations... 142
    Simulation Case Study: The Redesign of the Crime Arrest
    Process at a UK Police Force 143
    Discussion 144
Summary 145
References 146

Index 151

1

Introduction to Simulation

Introduction

This chapter aims to introduce the technique of simulation modelling and define the type of simulation discussed in this text. The relevance of simulation to organisations is explained by reference to the behaviour of variability and interdependence shown by processes within organisational systems. The potential benefits and disadvantages of using simulation in the organisation are then discussed.

What is Simulation Modelling?

Simulation is used to mean a number of things from a physical prototype to a video game. As it is used in this text simulation refers to the use of a model to investigate the behaviour of a business system. The use of a model on a computer to mimic the operation of a business means that the performance of the business over an extended time period can be observed quickly and under a number of different scenarios. The simulation method usually refers to both the process of building a model and the conducting of experiments on that model. An experiment consists of repeatedly running the simulation for a time period in order to provide data for statistical analysis. An experiment is conducted in order to understand the behaviour of the model and to evaluate the effect of different input levels on specified performance measures. Pidd (1996) characterises systems best suited to simulation as:

Dynamic   their behaviour varies over time
Interactive   they consist of a number of components which interact with each other
Complicated   the system consist of many interacting and dynamic objects

Most organisational systems have these characteristics and thus simulation would seem to be an ideal tool for providing information on the behaviour of an organisation. In practice simulation is most widely used and appropriate for applications which involve queuing – of people, materials or information. By simply defining in the simulation the timing of arrival to the queue and the availability of the resource that is being queued for, then the simulation is able to provide performance statistics on the average time in the queue and the average queue size for a particular system. A simple example would be to determine the performance of a supermarket check-out system. From information provided on customer arrival rates and check-out service times the simulation would be able to report performance measures such as average customer queue times and the utilisation of the check-out resource. Queuing systems are prevalent and examples include raw material waiting for processing in a manufacturing plant, vehicle queuing in transportation systems, documents waiting for processing in a workflow system, patients waiting to be seen in a doctor's surgery and many others.

Simulation in general covers a large area of interest and in order to clarify the particular area of interest in this text a short explanation is given of common terms in this area. Simulation can refer to a range of model types from spreadsheet models, system dynamic simulations and discrete-event simulation. *Discrete-Event Simulation* or *Simulation Modelling* is the subject of this text. Early simulation systems generated reports of system performance, but advances in software and hardware allowed the development of animation capabilities. When combined with the ability to interact with the model this technique became known as *Visual Interactive Simulation (VIS)*. Most simulation modelling software is now implemented using graphical user interfaces using objects or icons that are placed on the screen to produce a model. These are often referred to as *Visual Interactive Modelling (VIM)* systems. The software used in this text is VIM's that use the discrete-event method of operation. Finally because of the use of simulation in the context of Business Process Reengineering (BPR) and of other process based change methods the technique is also referred to as *Business Process Simulation (BPS)*. The term *Business Process Modelling (BPM)* is also sometimes used, but this term is traditionally related to information system development tools.

Simulation and Variability

The use of business analysis techniques such as flow charts and spreadsheets is widespread and well established. However these techniques are unable to capture the range of behaviour of a typical process due to their inability to incorporate dynamic (i.e. time dependent) behaviour. There are two aspects of dynamic systems which need to be addressed:

Variability

Most business systems contain variability in both the demand on the system (e.g. customer arrivals) and in durations (e.g. customer service times) of activities within the system. The use of fixed (e.g. average) values will provide some indication of performance, but simulation permits the incorporation of statistical distributions and thus provides an indication of both the range and variability of the performance of the system. This is important in customer based systems when not only is the average performance relevant, but performance should not drop below a certain level (e.g. customer service time) or customers will be lost. In service systems, two widely used performance measures are an estimate of the maximum queuing time for customers and the utilisation (i.e. percentage time occupied) for the staff serving the customer.

Interdependence

Most systems contain a number of decision points that affect the overall performance of the system. The simulation technique can incorporate statistical distributions to model the likely decision options taken. Also the 'knock-on' effect of many interdependent decisions over time can be assessed using the simulation's ability to show system behaviour over a time period.

To show the effect of variability on systems, a simple example will be presented. A manager of a small shop wishes to predict how long customers wait for service during a typical day. The owner has identified 2 types of customer, who have different amounts of shopping and so take different amounts of time to serve. Type A customers account for 70% of custom and take on average 10 minutes to serve. Type B customers account for 30% of custom and take on average 5 minutes to serve. The owner has estimated that during an 8 hour day, on average the shop will serve 40 customers. The owner then calculates the serve time during a particular day:

Customer A = 0.7 x 40 x 10 minutes = 280 minutes
Customer B = 0.3 x 40 x 5 minutes = 60 minutes
Therefore the total service time = 340 minutes and gives a utilisation of the shop till of 340/480 x 100 = 71%

Thus the owner is confident all customers can be served promptly during a typical day. A simulation model was constructed of this system to estimate the service time for customers. Using a fixed time between customer arrivals of 480/40 = 12 minutes and with a 70% probability of a 10 minutes service time and a 30% probability of a 5 minutes service time, the overall service time for customers has a range of between 5 to 10 minutes and no queues are present in this system.

Service Time for Customer (minutes)
Average   8.5
Minimum   5
Maximum   10

However in reality customers will not arrive, equally spaced at 12 minute intervals, but will arrive randomly with an average interval of 12 minutes. The simulation is altered to show a time between arrivals following an exponential distribution (the exponential distribution is often used to mimic the behaviour of customer arrivals) with a mean of 12 minutes. The owner was surprised by the simulation results:

Service Time for Customer (minutes)
Average   17
Minimum   5
Maximum   46

The average service time for a customer had doubled to 17 minutes, with a maximum of 46 minutes! The example demonstrates how the performance of even simple systems can be affected by randomness. Variability would also be present in this system in other areas such as customer service times and the mix of customer types over time. The simulation method is able to incorporate all of these sources of variability to provide a more realistic picture of system performance.

Why Use Simulation?

Simulation Modelling is used to assist decision-making by providing a tool that allows the current behaviour of a system to be analysed and understood. It is also able to help predict the performance of that system under a number of scenarios determined by the decision maker. Simulation Modelling is useful in providing the following assistance to the process improvement effort:

- *Allows Prediction* – Predicts business system performance under a range of scenarios
- *Stimulates Creativity* – Helps creativity by allowing many different decision options to be tried quickly and cheaply.
- *Avoids Disruption* Allows a evaluation of a number of decision options without disruption or use of a real system
- *Reduces Risk* – Allows the evaluation of a number of possible scenario outcomes, permitting contingencies to be formulated for these outcomes and therefore reducing the risk of failure.
- *Provides Performance Measures* – Can be integrated into performance measurement systems to provide organisational performance measures and cost estimates.
- *Acts as a Communication Tool* – The results and computer animation can provide a forum for understanding the system behaviour. The dynamics of a system can be visualised over time aiding understanding of system interactions.
- *Assist Acceptance of Change* – Individuals can predict the effects of change, thus allowing them to accept and understand change and improve confidence towards implementation.
- *Encourages Data Collection* – The systematic collection of data from a variety of sources necessary to build the model, in itself can lead to new

insights on the operation of the system, before the model has been built or experimentation begun.
- *Allows Overview of Whole Process Performance* – Using simulation to model processes across departmental boundaries allows improvement of the whole process, rather than the optimisation of local activities at the expense of overall performance.
- *Training Tool* – Allows personnel to be trained or provides a demonstration of process behaviour without the possible cost and disruption to the real system.
- *Design Aid* – Allows process behaviour to be observed and thus optimised at an early stage in the process design effort.

Disadvantages of the Simulation Method

Although simulation can be applied widely in the organisation a model developed for a non-trivial problem will consume a significant amount of resource in terms of staff time. Both time and cost elements need to be considered. Thus an assessment must be made of costs against potential benefits. As with many investment decisions however the costs are usually substantially easier to estimate than potential benefits which may be of a more intangible nature. For example a benefit of greater staff knowledge which may lead to increased productivity. Because of the significant cost of a simulation analysis it is also important to consider alternative modelling methods which may provide the necessary information. These include such tools as spreadsheet analysis, queuing theory and linear programming. It is important to be aware however that although these tools may provide a quicker 'decision', approaches such as queuing theory makes a number of assumptions about the system being studied which can provide an inaccurate analysis. The importance of the ability of simulation to model the variability characteristics of a particular system should be carefully considered in these cases. Although simulation can study more complex systems than many analytical techniques, its use may be of limited value for very complex or unpredictable systems. For example human-based systems with staff which have discretion in their duties and how they undertake them present a particular challenge.

Even if a cost-benefit analysis has been made in favour of simulation, a factor that can discount the approach is insufficient time available to complete the project. Activities such as data collection and model building may take longer than is available before a decision is required. The best policy is to consider the use of simulation at an early stage in the decision process. A possible solution is to employ consultants or simulation experts who can reduce the project duration by employing additional staff and can provide a faster model build through the knowledge gained from previous projects.

Summary

This text covers the area of the use of discrete-event simulation in the analysis of organisational systems. It is relevant to organisations because its analysis can incorporate the attributes of variability and interdependence that organisations exhibit. It is able to help predict behaviour and so reduce risk while avoiding disruption to organisation systems. Disadvantages of the simulation approach include the time taken to conduct the study and the skills required to undertake the analysis.

References

Pidd, M. (1996) Tools for Thinking: Modelling in Management Science, Wiley, New York, NY.

2

The Usage of Simulation

Introduction

This chapter provides evidence of the current usage of the simulation technique in organisations. This is determined from evidence from published studies and a study undertaken by the author. There follows an overview of application areas of simulation in organisations that do use the technique. Applications of simulation in both manufacturing and service industry are provided. The widespread use of simulation in people-based applications has lead to an increasing interest in the modelling of human behaviour in simulation. A range of approaches to meeting this challenge are presented.

Published Surveys of the Use of Simulation

A number of surveys of simulation use have been conducted. In the USA Christy and Watson (1983) conducted a survey on the application of simulation in industry to members of the Institute of Management Sciences. It found that 89% of the firms that responded used the simulation technique, of which 32% initiated its use between 1975 and 1979. Kirkpatrick and Bell (1989) conducted a survey of visual interactive modelling use by conducting users through the simulation software supplier organisations. System complexity was the most cited reason for using VIM. Groups developing a VIM capability reported an expanding demand for their services and a growth in their prestige and credibility with management. Cochran et al. (1995) conducted a survey to provide information on simulation practices in industrial settings. The report finds that the simulation languages are being increasingly being discarded for simulators (VIMs). Somewhat optimistically the authors predicted that the ease of use and built-in methodology of simulators will mean opportunities for simulation consulting may die out. McLean and Leong (2001) also characterise the implementation of simulation in the US as 'limited' and 'sporadic'. Further studies indicate that the use of simulation is widespread or always successfully implemented in industrial settings (Upton, 1995; Beach et al., 2000).

The first major survey in the UK was by the Simulation Study Group (1991), the results of which can also be found in Hollocks (1992). This found that awareness of the technique in manufacturing industry as very low and some £300 million of benefits were being missed. The survey revealed only 11% of UK manufacturers using simulation as a decision support tool. Other than awareness, the principal obstacles to wider use were found to be lack of skills or training, difficulties in data acquisition, the time taken to build the model and a lack of management commitment. The ESPIRIT working group on Simulation in Europe (SiE) stated that this general picture of proliferation was reflected across Europe (Kerckhoffs et al., 1995). Hollocks (2001) maintained that there has been no evidence to indicate any material shift in the level of simulation use on UK manufacturing since the original survey. Edge and Klein (1993) conducted a survey to assess use of operational research techniques in small businesses in the UK. Out of the 19 firms surveyed, 15 had no familiarity with simulation, 4 had moderate familiarity and 0 had high familiarity. Wisniewski et al. (1994) conducted an investigation into the use of quantitative techniques by business in Denmark, Scotland and the UK. It found 57% of companies surveyed had awareness of simulation and of those 63% actually used the technique. The major factor cited for non-use of a technique was that the firm did not see the technique as being relevant to the organisation's activities. A survey covering non-academic members of INFORMS found only 10% of survey respondents employed simulation as a decision support tool (Abdel-Malek et al., 1999).

Hlupic reports on a study undertaken in 1997 in Hlupic (1999) and Hlupic (2000) on current academic and industrial users of simulation. The report finds that two-thirds of respondents use simulation for manufacturing applications. Other significant application areas are health (18.5%) and communication systems (11.1%). The survey focuses on simulation software and finds that the most requested software features are more assistance in experimental design, packages that are easier to learn and use and improved software compatibility. A survey by Fryer and Garnett (1999) considers the issue of a lack of education and training as a barrier to the wider and more effective use of simulation. They found that many courses within higher education place emphasis exclusively in theoretical and technical issues, and ignore practical skills, which may not be appropriate for modelling relatively simple business processes. Thus the challenge for the simulation education community is to find the right balance between theoretical, technical and practical illustration. Melao and Pidd (2003) conducted a survey among potential business process simulation (BPS) users. BPS is defined as the use of a computer simulation model to mimic a business process so as to consider changes before their implementation. The aim of the survey was to understand the requirements of people engaged in this relatively new area of simulation usage. The survey revealed a low usage in the design, modification and improvement of business processes. There was no evidence of a skills gap found, but rather a feeling that there is no net gain from employing simulation methods when simpler methods will suffice.

A Survey of Simulation Use in the UK

A descriptive survey is presented of the current usage of simulation in UK organisations. The survey sampled approximately 1000 companies from a database of UK organisations across industrial sectors. The survey was designed to gather data using a Likert scale in response to a number of questions regarding the use of the simulation technique. Space is also provided for comments for qualitative feedback. The response to this questionnaire was very low and so the question of non-response bias needs to be addressed. However it was felt likely that the majority of non-responses were non-users of simulation and so the proportion of non-users is greatly understated. From the survey 30 responses were received giving a response rate of 3%. Of these 24 (80%) were from non-users of simulation and 6 (20%) were from users of the technique. The non-users of simulation were asked a number of questions to identify the main reasons why the simulation technique is not used at this time. Users were requested to respond on a Likert scale numbered 0 (not a reason) to 4 (major reason) to questions categorised under the headings of cost, awareness, skills, organisational and limitations of technique. The following analysis was performed on the responses in order to provide a summary of the results. Responses of 0 and 1 on the scale were combined and converted to a percentage of the sample size, responses of 2 on the scale were discarded, and responses of 3 and 4 were combined and converted to a percentage of the sample size. A summary of the responses is provided in Table 2.1.

Table 2.1. Survey Results of 'Reasons for non-use of Simulation'

Reasons for non-use of simulation	Major Reason %	Not a Reason %
Costs		
Initial cost of software, hardware and training	20.83	33.33
Project costs of model developer time	12.5	33.33
Project costs of staff time in data collection	12.5	37.5
Cost of simulation consultancy	20.83	29.17
Awareness		
Lack of knowledge of where technique can be applied	**54.17**	25
Lack of knowledge of benefits of technique	**50**	20.83
Assumption that technique is only used in manufacturing applications	8.33	37.5
Skills		
Lack of staff skills in model building	33.33	25
Lack of staff skills in statistical analysis	20.83	37.5
Organisational		
Lack of confidence that results of a simulation study will actually lead to change	25	**41.67**
Study results takes discretion away from decision makers	8.33	**41.67**
Lack of understanding of technique by decision makers	25	25
Intuitive (non analytical) approach to decision making preferred	25	25

Table 2.1. (continued)

Limitations of Technique		
Lack of ability of simulation to model human-based or ill-defined processes	20.83	25
Other techniques (e.g. spreadsheet modelling) provide sufficient information	**50**	12.5
Time taken to develop model and present results too slow for decision making	16.67	20.83
Lack of flexibility in reusing models	12.5	33.33
Lack of ability to run models in real-time	4.17	**45.83**
Assumptions made in model make it an unrealistic representation of the real system	16.67	29.17
Lack of connectivity of simulation with databases and IT systems	12.5	37.5

It can be seen from Table 2.1 that the most likely reason for non-use of simulation is a lack of awareness of the technique. This is found both in a lack of knowledge of where the technique can be applied and a lack of knowledge of the benefits of the technique. The other major reason given for non-use of simulation is the view that other techniques, such as spreadsheet modelling, provides sufficient information for decision making purposes. The lack of ability to run models in real-time, a lack of confidence that results of a simulation study will actually lead to change and a feeling that study results takes discretion away from decision makers were not felt to be reasons for non-use by a significant proportion of respondents. A space was provided for a qualitative response to how the respondents felt the simulation technique could be used more widely. Responses included the following:

'Whilst in theory a good idea, in the real world it is only appropriate for large companies with complex procedures'

'Most of our problems are not complex enough to need simulation modelling'

'I have never HEARD of Simulation Modelling'

'Simulation techniques are tools used by half witted management consultants who don't understand business'

'As an ISO 9001 QA service company we track and measure all processes using D/B and s/sheets. Is cost-effective and meets our needs'

These responses point to a lack of awareness of the technique and its perceived non-applicability to companies. The survey also requested responses for current users of simulation. In this case the users of simulation were asked the same questions as non-users but in the context of why greater use was not made of simulation in their organisation. The responses are presented in Table 2.2.

Table 2.2. Survey Results of 'Major barrier to greater use of simulation?'

Major Barrier to greater use of Simulation?	Major Barrier %	Not a Barrier %
Costs		
Initial cost of software, hardware and training	50	16.67
Project costs of model developer time	50	16.67
Project costs of staff time in data collection	**83.33**	16.67
Cost of simulation consultancy	**83.33**	16.67
Awareness		
Lack of knowledge of where technique can be applied	50	50
Lack of knowledge of benefits of technique	33.33	50
Assumption that technique is only used in manufacturing applications	16.67	66.67
Skills		
Lack of staff skills in model building	33.33	16.67
Lack of staff skills in statistical analysis	33.33	0
Organisational		
Lack of confidence that results of a simulation study will actually lead to change	16.67	66.67
Study results takes discretion away from decision makers	33.33	66.67
Lack of understanding of technique by decision makers	16.67	66.67
Intuitive (non analytical) approach to decision making preferred	0	33.33
Limitations of Technique		
Lack of ability of simulation to model human-based or ill-defined processes	16.67	33.33
Other techniques (e.g. spreadsheet modelling) provide sufficient information	33.33	33.33
Time taken to develop model and present results too slow for decision making	16.67	33.33
Lack of flexibility in reusing models	50	33.33
Lack of ability to run models in real-time	16.67	66.67
Assumptions made in model make it an unrealistic representation of the real system	16.67	66.67
Lack of connectivity of simulation with databases and IT systems	33.33	16.67

It would seem from the results that cost is a significant barrier to greater use of the technique. Cost is seen as significant both in terms of staff time required for data collection and the cost of consultancy for outsourced simulation development. Overall the survey results show that while there is scepticism of the applicability of the technique in non-users, actual users of the technique seem to be satisfied with the relevance of the technique, with the main barrier to further use being cost. This suggests that the factors affecting the breadth of use across organisations differ from the factors affecting the depth of use within an organisation. Although difficult to judge from the small response rate, it seems that this survey confirms a low usage of simulation by organisations reported in the literature search.

Where in the Organisation is Simulation Used?

Simulation Modelling is used in various areas of many different types of organisations. Some examples of simulation use are given below, with reference to industrial case studies published by the author.

Manufacturing Applications

In order to remain competitive manufacturing organisations must ensure their systems can meet changing market needs in terms of product mix and capacity levels whilst achieving efficient use of resources. Because of the complex nature of these systems with many interdependent parts, simulation is used extensively to optimise performance. Greasley (2004) provides a case study of using simulation to optimise a line type manufacturing system by ensuring that each stage in the line has an equal capacity level. Greasley (2000c) provides a case study of the use of simulation to test various scheduling scenarios on a just-in-time production system in order to optimise resource utilisation. Greasley (2005) provides a case study of the use of simulation to develop production rules that would reduce work-in-progress and production lead-time.

For large capital investments such as equipment and plant, simulation can reduce the risk of implementation at a relatively small cost. Simulation is used to ensure the equipment levels and plant layout is suitable for the planned capacity requirements of the facility.

A key customer requirement of any delivered manufactured good or service supplied is its reliability in operation which is often a key measure of service quality. Simulation can test the performance of a system under a number of scenarios both relatively quickly and cheaply. Steps can be then be taken in advance to ensure service is maintained under various operating conditions. Greasley (2000a) provides a case study of the use of simulation to ensure that a transportation system can meet demand within a planned maintenance schedule under a variety of scenarios of train breakdown events. Although simulation has traditionally been associated with improving internal efficiency of systems, this is a good example of where it was used not only to prove capability to the maintenance operator, but also to provide a tool to demonstrate to the potential customer that performance targets could be met.

Transportation systems such as rail and airline services as well as internal systems such as automated guided vehicles (AGVs) can be analysed using simulation. Many simulation software packages have special facilities to model track-based and conveyor type systems and simulation is ideally suited to analyse the complex interactions and knock-on effects that can occur in these systems.

Service Applications

The productivity of service sector systems has not increased at the rate of manufacturing systems and as the service sector has increased the potential increase in productivity from improving services has been recognised. The use of BPR and other methodologies to streamline service processes have many parallels in techniques used in manufacturing for many years. Simulation is now being used to help analyse many service processes to improve customer service and reduce cost. Greasley (2003) provide a case study of the use of simulation to analyse a proposed workflow system for a group of estate-agency outlets. The purpose of the workflow system is to automate the paper-flow in the house-buying process and thus increase speed of service to the customer. The simulation was used to assist in predicting demand levels on the system, identifying bottlenecks and thus optimise operation before implementation.

Business Process Reengineering (BPR) attempts to improve organisational performance by analysis of a business from a process rather than a functional perspective and then redesign these processes to optimise performance. Greasley and Barlow (1998) provide a case study of the use of simulation in the context of a BPR project to redesign the custody operation in a UK Police Service. Greasley (2000b) and Greasley (2001) present a case study showing how simulation can be used in conjunction with the technique of activity-based-costing (ABC) to show costs from a cost, resource and activity perspective. The cases show how simulation can be used during most stages of a BPR initiative and how the use of simulation can be prioritised and aligned with strategic objectives through the use of techniques such as the balanced scorecard. They also show the ability of simulation to provide a variety of performance measures such as utilisation of people, speed of service delivery and activity cost.

The emphasis on performance measures in government services such as health care has led to the increased use of simulation to analyse systems and provide measures of performance under different configurations. Simulation is used to predict the performance of the computerisation of processes. This analysis can include both the process performance and the technical performance of the computer network itself, often using specialist network simulation software.

Modelling Human Behaviour with Simulation

The issue of how to incorporate human behaviour in a simulation study has become prominent due to the increased use of the technique in areas where human behaviour has an impact on process performance. These areas include business process initiatives, particularly those in the service industries and the increase in professional service applications, as opposed to mass manufacturing applications dominated by equipment and machinery. However despite the continued use of simulation in organisations and a relatively large skill base of simulation practitioners, the use of the technique to model human behaviour is not widespread. This section suggests a number of approaches to this task within the process of the simulation study. The different approaches have been derived from a literature review and categorised by level of abstraction. The aim is to raise awareness amongst simulation practitioners of the options available for incorporating human behaviour in their studies, show examples of the implementation of each method and provide guidance on when each method should be used.

Methods of Modelling Human Behaviour

From a literature review potential methods of modelling people were identified and classified by the level of abstraction of human behaviour (see Figure 2.1). A classification by level of abstraction (i.e. the level of detail of the modelling of human behaviour) was used because this was identified has having a major impact on the approach used to model human behaviour. Each approach is given a method name and listed in order of the level of abstraction used to model human behaviour. The framework recognises that the incorporation of human behaviour in a simulation study does not necessarily involve the coding of human behaviour in the simulation model itself. Thus the methods are classified into those that are undertaken 'outside' the model, and those that actually involve changes to the

Method Name	Method Description	World View	Model Abstraction	Simulation Method	Abstraction
Simplify	Eliminate human behaviour by simplification			None	Outside the Model
Externalise	Incorporate human behaviour outside of the model			Gaming Expert Systems Neural Networks	
Flow	Model humans as flows	Continuous	Macro	System Dynamics	Inside the Model
Entity	Model human as a machine or material	Process	Meso	Discrete Event Simulation	
Task	Model human action				
Individual	Model individual human behaviour	Object	Micro	Agent-Based Simulation	

Figure 2.1. Methods of modelling human behaviour in a simulation study

construction of the model ('inside' the model). Methods inside the model are classified in terms of world view and model abstraction in order to clarify the different levels of abstraction within this category. The framework then provides a suggested simulation method for each of the levels of abstraction identified from the literature.

The methods shown in Figure 2.1 are now described in more detail.

The simplify method involves the simplification of the simulation model in order to eliminate any requirement to codify human behaviour. This strategy is relevant because a simulation model is not a copy of reality and should only include those elements necessary to meet the study objectives. This may make the incorporation of human behaviour unnecessary. Actual mechanisms for the simplification of reality in a simulation model can be classified into omission, aggregation and substitution (Pegden et al., 1995). In terms of modelling human behaviour this can relate to the following:

Omission: omitting human behaviour from the model, such as unexpected absences through sickness. It may be assumed in the model that alternative staffing is allocated by managers. Often machine-based processes are modelled without reference to the human operator they employ.

Aggregation: processes or the work of whole departments may be aggregated if their internal working is not the focus of the simulation study.

Substitution: Human processes may be substituted by a 'delay' element with a constant process time in a simulation model thus removing any complicating factors of human behaviour.

The externalise approach attempts to incorporate human behaviour in the study, but not within the simulation model itself. The area of gaming simulation represents a specialist area of simulation when the model is being used in effect to collect data from a human in real-time and react to this information. Alternative techniques such as expert systems and neural networks can be interfaced with the simulation and be used to provide a suitable repository for human behaviour. There will however be most likely a large overhead in terms of integrating these systems with simulation software.

The flow method models humans at the highest level of abstraction using differential equations. Continuous modelling has been used for many years and has been integrated in discrete-event simulation software packages such as SIMAN/CINEMA (Pegden at al, 1995) and its replacement ARENA (Kelton et al., 2007). The implementation of the continuous world view has however been usually associated with the use of the system dynamics technique (Forrester, 1961). A benefit of the system dynamics approach is that it provides a way of understanding the underlying structure of a human system in order to explain behaviour. The level of abstraction however means that system dynamic models do not possess the ability to carry information about each entity (person) through the system being modelled and are not able to show queuing behaviour of people derived from demand and supply (Stahl, 1995). Thus the simulation of human behaviour in customer processing applications for example may not be feasible using this approach.

The most basic way to model human behaviour using the process world view is to either represent people by simulated machines (resources) and/or simulated materials (entities). This allows the availability of staff to be monitored in the case of resources and the flow characteristics of people, such as customers, to be monitored in the case of entities. This will provide useful information in many instances, but does not reflect the way people actually work, particularly in a service context where their day-to-day schedule may be a matter of personal preference.

The task approach attempts to model human behaviour without the complexity of modelling the cognitive and other variables that lead to that behaviour. The rationale behind the approach is described by Shaw and Pritchett (2005), 'In this approach models are described as modelling performance rather than behaviour because of their scope – the current state of the art is better at capturing purposeful actions of a human as generated by well-understood psychological phenomenon, than it is at modelling in detail all aspects of human behaviour not driven by purpose.' Laughery (1999) models human performance by breaking down activities into a number of tasks. This method is intended to get closer to human performance than resource modelling by being able to model the individual actions that humans do rather than subsuming many parallel tasks it one generic task. Elliman et al. (2005) uses task and environmental variables, rather than individual characteristics to model individual behaviour. Thus provides a simpler method than using individual differences but still requires the need to operationalise the variables chosen. Bernhard and Schilling (1997) model people using the entity method but separate material flow from people flow. No individual differences are taken into account and the approach uses a centralised mechanism/database to control workers. These examples show the DES method is a suitable platform for modelling humans using a task approach.

The individual approach attempts to model the internal cognitive processes that lead to human behaviour. This has the advantage of not necessarily being task specific, but requires a sophisticated model of human behaviour. The difficulty of implementation of studies on human behaviour by behavioural and cognitive researchers is a significant barrier to this approach. Silverman et al. (2003) state 'there are well over one million pages of peer-reviewed, published studies on human behaviour and performance as a function of demographics, personality differences, cognitive style, situational and emotive variables, task elements, group and organisational dynamics and culture'. However Silverman (1991) states 'unfortunately, almost none of the existing literature addresses how to interpret and translate reported findings as principles and methods suitable for implementation or synthetic agent development'. Another barrier is the issue of the context of the behaviour represented in the simulation. Silverman states 'many first principle models from the behavioural science literature have been derived within a particular setting, whereas simulation developers may wish to deploy these models in different contexts'. Another barrier is that validation of multiple factors of human behaviour is difficult when the research literature is largely limited to the study of the independent rather than the interactive effects of these factors. Silverman et al. (2003) calls for behavioural scientists to provide a fuller representation of factors that influence human performance. Despite these barriers a number of architectures that model human cognition, such as PECS (Schmidt, 2000), Soar (Newell, 1990) and TPB (Ajzen, 1991) are available. Agent based systems are particularly useful when attempting to model autonomous human behaviour which is difficult in a DES, such as decision making and free physical movement around their workplace. There has been work on integrating DES and Agent based models, for example Dubiel and Tsimhoni (2005). This study combines an agent-based modelling of the free movement and interaction of people in a theme park with the discrete-event implementation of a theme park ride.

Choosing a Method to Simulate Human Behaviour

When considering whether to incorporate human behaviour in a simulation study, the major determinant should be the level of abstraction required to meet the study objectives. It may be that a simplification approach may be sufficient or human behaviour can be externalised to data structures such as an expert system. When considering incorporating human behaviour within the simulation there is an association of the technique of system dynamics at a macro level, discrete-event simulation at a meso level and agent-based simulation at a micro level. However these three techniques are capable of modelling at all three levels of abstraction and the skill-set of the modellers, who tend to operate using only one of the three world views may override the categorisation provided. There is also the issue of applications which require abstraction at the micro, meso and macro level within a single model. Ni (2006) provides an example of a traffic simulation which requires traffic flows to be modelled as flows across large distances (macro), as individual vehicles with characteristics (meso) and as autonomous systems modelling driver cognition (micro). The ideal here would be simulation software that provides tools representing all 3 world views and these do exist, for example AnyLogic (www.xjtek.com). Pegden (2005) predicts that the next generation simulation tool will combine the continuous, process and object world views by making the terms model and object interchangeable, so model builders become object builders.

Summary

The surveys confirm a low rate of usage of the simulation technique, partly due to the perceived cost of establishing a simulation capability in the organisation. A number of applications of simulation in both manufacturing and service industries are provided. A framework is presented providing a range of approaches to modelling human behaviour in a simulation model. The framework suggests a modelling approach according to the level of abstraction of human behaviour appropriate to meet the simulation study objectives.

References

Abdel-Malek, L.; Johnson, F. and Spencer, III, T. (1999), OR practice: Survey results and reflections of practising informs members, *Journal of the Operational Research Society*, 50: 10, 994–1003.

Ajzen, A. (1991), The theory of planned behaviour, *Organizational Behavior and Human Decision Processes*, 50, 179–211.

Beach, R.; Muhlemann, A.P.; Price, D.H.R. and Sharp, J.A. (2000), The selection of information systems for production management: an evolving problem, *International Journal of Production Economics*, 64, 319–329.

Bernhard, W. and Schilling, A., (1997), Simulation of Group Work Processes in Manufacturing, *Proceedings of the 1997 Winter Simulation Conference*, SCS, 888–891.

Christy, D.P. and Watson, H. (1983), The application of simulation: A survey of industry practice, *INTERFACES*, 13, 47–52.

Cochran, J.K.; MacKulak, G.T.; Savory, P.A. (1995), Simulation project characteristics in industrial settings, *INTERFACES*, 25, 104–113.

Dubiel, B. and Tsimhoni, O., (2005), Integrating Agent Based Modeling into a Discrete Event Simulation, *Proceedings of the 2005 Winter Simulation Conference*, SCS, 1029–1037.

Edge, C. and Klein, J. (1993), No job too small?, *OR Insight*, 6:3, 8–11.

Elliman, T., Eatock, J. and Spencer, N., (2005), Modelling knowledge worker behaviour in business process studies, *The Journal of Enterprise Information Management*, 18: 1, 79–94.

Forrester, J., (1961), Industrial Dynamics, Productivity Press, Cambridge: MA.

Fryer, K.J. and Garnett, J. (1999) Simulation Education and Training: The Relationship between Higher Education and the Workplace, *Proceedings of the UKSIM '99 conference of the UK Simulation Society*, St. Catharine's College, Cambridge, ed. by Al-Dabass, D., Cheng, R., 134–140.

Greasley, A. (2000a), Using Simulation to Assess the Service Reliability of a Train Maintenance Depot, *Quality and Reliability Engineering International*, 16: 3, 221–228.

Greasley, A. (2000b), A Simulation Analysis of Arrest Costs, *Journal of the Operational Research Society*, 51, 162–167.

Greasley, A. (2000c), A Simulation of a Valve Manufacturing Plant, *Proceedings of the 2000 Summer Computer Simulation Conference,* Society for Computer Simulation, San Diego, USA.

Greasley, A. (2001), Costing Police Custody Operations, *Policing: An International Journal of Police Strategies & Management*, 24: 2, 216–227.

Greasley, A. (2003), A Simulation of a Workflow Management System, *Work Study*, 52: 5, 256–261.

Greasley, A. (2004), The case for the organisational use of simulation, *Journal of Manufacturing Technology Management*, 15: 7, 560–566.

Greasley, A. (2005), Using system dynamics in a discrete-event simulation study of a manufacturing plant, *International Journal of Operations and Production Management*, 25: 5/6, 534–548.

Greasley, A. and Barlow, S. (1998), Using Simulation Modelling for BPR: Resource Allocation in a Police Custody Process, *International Journal of Operations and Production Management*, 18: 9/10, 978–988.

Hlupic, V. (1999), Discrete-Event Simulation Software: What the Users Want, *SIMULATION*, 73, 362–370.

Hlupic, V. (2000), Simulation Software: A survey of academic and industrial users, *International Journal of Simulation*, 1: 1–2, 1–11.

Hollocks, B.W. (1992), A well kept secret? Simulation in manufacturing industry reviewed, *OR Insight*, 5: 4, 12–17.

Hollocks, B.W. (2001), Discrete event simulation: an inquiry into user practice, *Simulation Practice and Theory*, 8: 5, 451–471.

Kelton, W.D., Sadowski, R.P. and Sturrock, D.T., (2007), Simulation with Arena, 4th Edition, McGraw-Hill, NY.

Kerckhoffs, E.J.H.; Vangheluwe, H.L. and Vansteenkiste, G.C. (1975), Interim report on ESPRIT Basic Working Group 8467, http://biomath.rug.ac.be/~hv/SiE/ [Accessed 4 November 2002].

Kirkpatrick, P. and Bell, P.C. (1989), Visual interactive modelling in industry: Results from a survey of visual interactive model builders, *INTERFACES*, 19, 71–79.

Laughery, R., (1999), Using Discrete-Event Simulation to Model Human Performance in Complex Systems, *Proceedings of the 1999 Winter Simulation Conference*, SCS, 815–820.

McLean, C. and Leong, S. (2001), The expanding role of simulation in future manufacturing. In B.A. Peters, J.S. Smith, D.J. Medeiros and M.W. Rohrer (eds), *Proceedings of the 2001 Winter Simulation Conference* (Virginia), 1478–1486.

Melao, N. and Pidd, M. (2003), Use of Business Process Simulation: A survey of practitioners, *Journal of the Operational Research Society*, 54, 2–10.

Newell, A., (1990), Unified Theories of Cognition, Harvard University Press, Cambridge: MA.

Ni, D., (2006), A framework for new generation transportation simulation, *Proceedings of the 2006 Winter Simulation conference*, SCS, 1508–1514.

Pegden, C.D., Shannon, R.E. and Sadowski, R.P., (1995), Introduction to Simulation using SIMAN, 2nd Edition, McGraw-Hill.

Pegden, C.D., (2005), Future Directions in Simulation Modeling, *Proceedings of the 2005 Winter Simulation conference*, SCS.

Schmidt, B., (2000), The Modelling of Human Behaviour, SCS Publications.

Shaw, A.P. and Pritchett, R., (2005), Agent-Based Modeling and Simulation of Socio-Technical Systems in Rouse, W.B. and Boff, K.R. (eds) *Organizational Simulation*, John Wiley & Sons, Inc.

Silverman, B.G., Cornwell, J. and O'Brien, K., (2003), Human Performance Simulation in Ness, J.W., Rizer, D.R., & Tepe, V. (eds) *Metrics and methods in human performance research toward individual and small unit simulation*, Human Systems Information Analysis Centre: Washington DC.

Silverman, B.G., (1991), Expert critics: Operationalising the judgement/decision making literature as a theory of "bugs" and repair strategies, *Knowledge Acquisition*, 3, 175–214.

Simulation Study Group (1991), *Simulation in U.K. Manufacturing Industry*, R. Horrocks (ed.), The Management Consulting Group, University of Warwick Science Park.

Stahl, I., (1995), New Product Development: When Discrete Simulation is Preferable to System Dynamics, *Proceedings of the 1995 EUROSIM Conference*, Elsevier Science B.V.

Upton, D.M. (1995), What makes factories flexible, *Harvard Business Review*, 73: 4, 74–84.

Wisniewski, M.; Jones, C.; Kristensen, K.; Madsen, H.; Ostergaard, P. (1994), Does anyone use the techniques we teach?, *OR Insight*, 7:2, 2–6.

3

Acquiring the Resources for Simulation

Introduction

The use of simulation is both a technical issue involving model development and analysis and a process of the implementation of organisational change. This chapter discusses technical issues such as the selection of simulation software and organisational issues such as the selection of personnel and the acquisition of resources required to provide the capability to undertake a simulation project. It is important that the full costs of introducing simulation are considered, including user time and any necessary training activities. The potential benefits of simulation must also be estimated. One of the reasons simulation is not used more widely is the benefits from change, undertaken as a result of a simulation study can be difficult to quantify.

Steps in Introducing Simulation

The steps in introducing simulation in the organisation are outlined.

1. Select Simulation Sponsor

If the organisation has not utilised the simulation method previously then it may be necessary to assign a person with responsibility for investigating the relevance and feasibility of the approach. This person will ideally have both managerial understanding of the process change that simulation can facilitate and knowledge of data collection and statistical interpretation issues which are required for successful analysis. The development of training schemes for relevant personnel should be investigated so the required mix of skills and experience is present before a project is commenced. It may be necessary to use consultancy experience to guide staff and transfer skills in initial simulation projects.

2. Evaluate Potential Benefits of Simulation

Often the use of simulation modelling can be justified by the benefits accruing from a single project. However due to the potentially high setup costs in terms of the purchase of simulation software and user training needs the organisation may wish to evaluate the long-term benefits of the technique across a number of potential projects before committing resources to the approach. This assessment would involve the simulation project sponsor and relevant personnel in assessing potential application areas and covering the following points

- Do potential application areas contain the variability and time dependent factors which make simulation a suitable analysis tool?
- Do the number and importance of the application areas warrant the investment in the simulation technique?
- Is there existing or potential staff expertise and support to implement the technique?
- Are sufficient funds available for aspects such as software, hardware, and training and user time?
- Is suitable simulation software available which will enable the required skills to be obtained by staff within a suitable timeframe?
- Will sufficient management support in the relevant business areas be forthcoming in the areas of the supply of data and implementation of changes suggested by the technique?
- Are there opportunities for integration with other process improvement tools such as Process Mapping, Activity Based Costing (ABC) and Workflow Management Systems (WFMS)?
- Does the level of uncertainty/risk in change projects, increase the usefulness of simulation as a technique to accept change and increase confidence in implementing new practices?

3. Estimate Resource Requirements

The main areas to consider in terms of resource requirements when implementing simulation are as follows:

Software
Most simulation software has an initial cost for the package and an additional cost for an annual maintenance contract which supplies telephone support and upgrades. It is important to ensure that the latest version of the software is utilised as changes in software functionality can substantially enhance the usability of the software and so reduce the amount of user development time required. Most software is available for the Windows platform, but packages are also available for UNIX and other systems.

Hardware
Assuming a dedicated computer is required; costs for a suitable machine should be included. The hardware specification should match the requirements of the simulation software chosen. Maintenance and replacement costs of hardware should be included in cost estimates.

Training
Initial training will speed future software development. Training may be required in both statistical techniques and model development in the software package chosen. Training may be provided by the software supplier, Universities or consultancy organisations.

Staff Time
This will be the most expensive aspect of the simulation implementation and can be difficult to predict, especially if simulation personnel are shared with other projects. The developer time required will depend on both the experience of the person in developing simulation models, the complexity of the simulation project and the number of projects it is intended to undertake. Time estimates should also factor in the cost of the time of personnel involved in data collection and other activities in support of the simulation team.

4. Selecting the Simulation Software Type

Before the type of discrete-event simulation system is chosen it is important to ensure that the right type of modelling approach is used for the particular problems that may be encountered in the organisation. In general discrete-event simulation modelling can be seen to be appropriate for dynamic systems, i.e. to investigate systems that change over time, but if the decision maker can recognise the system as a static model a computer spreadsheet model may be the most appropriate modelling technique to use. Chapter 1 outlines the main types of simulation systems. If it is decided that discrete-event simulation is to be used then before a simulation software package is chosen it is necessary to choose the type of implementation required. The three main categories of simulation systems are as follows:

General Purpose Languages
Computer languages such a FORTRAN, C and C++ have been used for many years to construct simulation models. Davies and O'Keefe (1989) provide computer code for a simulation system using the BASIC and PASCAL computer programming languages. Pidd and Cassel (2000) provides code for a JAVA implementation. Although these languages enable the user maximum flexibility in model building they are now unlikely to be appropriate for business users of the technique due to the greatly increased development times they require compared to a specialised simulation package. A discrete-event spreadsheet simulation system has been developed by the author (Greasley, 1998) using the Visual Basic for Applications (VBA) language which is included with the Microsoft Excel spreadsheet. Although not intended for the analysis of complex systems it does

enable users to gain "hands-on" experience of using simulation on the familiar spreadsheet platform. The software also allows the user to see the internal workings of the discrete-event method such as the operation of the event calendar for scheduling future events.

Simulation Languages

Many dedicated simulation programming languages have been developed which hide the workings of the discrete-event system from the user and provide a specialised command language for constructing simulation models. These systems will also provide facilities for tracing the status of the model through time in order to aid model validation. Many systems also incorporate a graphical animated display of the simulation model. Computer languages developed specifically for constructing discrete-event simulation models include SIMAN, SIMSCRIPT, SLAM and GPSS. These provide a much quicker method of building simulation models than using a general purpose language but have potential disadvantages of the cost of purchasing the software and the time needed to learn the simulation language.

Visual Interactive Modelling (VIM) Systems

Visual Interactive Modelling systems include ARENA, WITNESS, SIMUL8 and SIMFACTORY. These packages are based on the use of graphic symbols or icons which reduce or eliminate the need to code the simulation model. A model is instead constructed by placing simulation icons on the screen which represent different elements of the model. For example a particular icon could represent a queue. Data is entered into the model by clicking with a mouse on the relevant icon to activate a screen input dialog box. Animation facilities are also incorporated into these packages. For most business applications a VIM system is the most appropriate, although the cost of the software package can be high. VIM's are basically a graphical 'front-end' to a simulation language which takes its information from the parameters of the icons placed on the screen rather than from coding direct. For example the ARENA system is based on the SIMAN simulation language. However these systems do not release the user from the task of understanding the building blocks of the simulation system or understanding statistical issues. Also a complex model may well require recourse to language constructs.

Discussion of Choice of Simulation Package

The following discusses the choice available to the user in selecting the most appropriate simulation software package. General purpose software may be applicable when funds are not available to purchase a specialist simulation package or the developer wishes to produce a model which can be distributed amongst a group of people. An example of this is the production of a simulation on a disk for a student cohort. For commercial applications the choice will be between Simulation languages and VIM's. The main strength of simulation languages being their relatively low cost and flexibility in the range of models they can produce. They do however have the disadvantage, as with any computer software language, of the need to train users and ensure that programming expertise is not lost when an employee leaves the organisation. Although VIM's usually provide a more restricted range of applications they have the advantage of shorter model build and validation times and are easier to learn. The extra expenditure on VIM's should be recouped through decreased staff costs if the software is used for multiple projects. VIM's are strongly recommended for use by non-technical users or when an end-user approach is taken. However detailed modelling may be difficult within the constraints associated with a VIM. Systems such as ARENA allow the user to incorporate simulation language commands within a VIM model. This increases flexibility, but also brings the disadvantages associated with a simulation language. However whichever simulation software tool is chosen it is likely that for any reasonably complex system that some programming will be required (Banks and Gibson, 1997). It must also be noted that using a VIM does not free the user from the need to conduct the statistical aspects of model construction and validation. One approach to package selection is to assess the package in terms of the needs of a decision support system in terms of the Dialog (interface facilities), Data (database integration) and Modelling (constructs) perspectives (Sprague, 1993).

For the dialog component the user dialogue must successfully balance the need for ease of use for what may be a casual computer user and flexibility in allowing a number of approaches to solving the problem. The two main dialog approaches are the graphical input system which provides a simple, controlled interface and a command language which provides more flexibility but requires more knowledge of the system. Most decision-makers are unwilling to be trained to use a complex interface and so the command language approach usually requires an intermediary to operate the system. The graphical input system offered by VIM systems is thus usually preferred. Other facilities that help usability include automated input and output analysis help and debugging features and a comprehensive output reporting facility. The visual capabilities of the package should also be taken into account, with features such as automatic animation provided and 3D modelling available. Most VIM systems consist of a draw package which allows a static background to be created using a number of draw and text objects. On this static background a dynamic element is created by defining a moving object shape which moves along a predefined path across the computer screen in response to changes in selected simulation status variables. Performance measures (e.g. queue length, time of customer/product in the system) can also be displayed and are automatically linked to defined simulation variables.

For the data component it is important that the simulation package can interface with other software such as spreadsheets and databases which can be used as a store for data used by the simulation. Simulation software has shown developments in this area recently with packages such as ARENA offering connectivity with the Microsoft Office family and the drawing package VISIO through the use of Visual Basic for Applications (VBA). Data should be available for import in a number of textual, logical or graphical formats such as ASCII, spreadsheet, ODBC, DLL and DXF. Numerical data can be used to provide timings for customer arrival or service times for example and graphical data, such as from a CAD package, can provide the static background graphics. Links to flowcharting software such as VISIO can be used to provide both the logical structure of the simulation as well as numerical data information.

For the model component the modelling component itself refers to the language used to construct the simulation model. A range of robust modelling functions is required to create (either from a probability distribution or from a user-defined schedule) and destroy entities (which can represent customers, products, information etc.) in the model. Functions are also required for modelling decision points which may operate using if-then or probability rules. Resource availability functions should be available which define an availability schedule and overlay on this non-availability due to random occurrences such as machine breakdown events. Functions should be available to assign multiple resources to a particular entity and a particular resource should be assignable to multiple entities in the system. An important aspect of the modelling functionality in a VIM is the ability to extend the standard functions available in the VIM by working at a lower level of programmability. Functionality may be increased further by allowing the user to access system variables and user-defined routines in the source code (usually C or C++) of the simulation software. Functions should also include both discrete and continuous probability distributions for modelling activity and arrival times and resource availability events. Software facilities which provide the 'best-fit' between a range of built-in distributions and user data are also required. Output analysis functions are also required in terms of providing relevant statistical information (e.g. average, standard deviation) over multiple runs of the simulation model.

5. Selecting the Simulation Software Package

Once a decision regarding the type of simulation software package (general purpose language, simulation language or visual interactive modelling system) has been made there needs to be a choice of which vendor to supply a particular simulation package. Examples of software in each category are shown in Table 3.1.

Table 3.1. Simulation Software Types

General Purpose language	Simulation language	Visual Interactive Modelling System
PASCAL	SIMAN	ARENA
C	GPSS	WITNESS
Visual Basic	GENETIK	SIMFACTORY
JAVA	SLAM II	PROMODEL

The potential user can read the software tutorial papers from the Winter Simulation Conference available on the internet site www.informs-cs.org/wscpapers.html which provides information of software available. Additional information can be obtained from both vendor representatives (especially a technical specification) and established users on the suitability of software for a particular application area. Vendors of simulation software can be rated on aspects such as:

- Quality of Vendor (current user base, revenue, length in business)
- Technical Support (type, responsiveness)
- Training (frequency, level, on-site availability)
- Modelling Services (e.g. consultancy experience)
- Cost of Ownership (upgrade policy, runtime license policy, multi-user policy)

A selection of simulation software supplier details is presented in Table 3.2.

Table 3.2. Simulation Software Vendors

Vendor	Software	Web Address
SIMUL8 Corporation	SIMUL8	www.simul8.com
Alion MA&D Operation	Micro Saint	www.maad.com/
ProModel Corporation	ProModel	www.promodel.com
Lanner Inc.	WITNESS	www.lanner.com
Rockwell Automation Inc.	ARENA	www.arenasimulation.com

Simulation software can be bought in a variety of forms including single-user copies and multi-user licenses. Most software has a price for the license and then as additional maintenance charge which covers telephone support and the supply of software upgrades. Some software allows 'run-time' models to be installed on unlicensed machines. This allows use of a completed model, with menu options that allows the selection of scenario parameters. Run-Time versions do not allow any changes to the model code or animation display however. It is also possible to obtain student versions (for class use in Universities) of software which contain all the features of the full licensed version by are limited in some way such as the size of the model or have disabled save or print functions. The software used within the case studies in this text is ARENA.

6. Computer Hardware Requirements

Most modern personal computers should be capable of running simulation software. Most software runs on a PC under WINDOWS (although software for other operating systems such as UNIX is available). The PC should have a relatively fast processor (e.g. Pentium) and adequate memory (e.g. 64MB) to allow suitable development and animation speeds. Once the particular package has been selected these details should be easily obtained from the vendor.

7. Training

To conduct a simulation modelling project successfully the project team should have skills in the following areas:

General Skills for all Stages of a Simulation Project

- Project Management (ensure project meets time, cost and quality criteria)
- Awareness of the application area. (e.g. knowledge of manufacturing techniques)
- Communication Skills (essential for definition of project objectives and data collection and implementation activities)

Skills Relevant to the Stages of the Simulation Study

- Data Collection (ability to collect detailed and accurate information)
- Process Analysis (ability to map organisational processes)
- Statistical Analysis (input and output data analysis)
- Model Building (simulation software translation)
- Model Validation (ability to critically evaluate model behaviour)
- Implementation (ability to ensure results of study are successfully implemented)

In many organisations it may be required that one person acquires all these skills. Rohrer and Banks (1998) emphasise the need to ensure that care is taken in not only choosing appropriate simulation software but in choosing people to undertake a simulation project that have the required skills. They categorise skills for simulation tasks into *required skills* which must be present before the individual can perform the task, *desired skills* which are optional for performance of the task and *acquired skills* which can be learned as the task is performed. Because of the wide ranging demands that will be made on the simulation analyst it may be necessary to conduct a number of pilot studies in order to identify suitable personnel before training needs are assessed.

Training is required in the steps in conducting a simulation modelling study as presented in this text, as well as training in the particular simulation software that is being used. Most software vendors offer training in their particular software package. If possible it is useful to be able to work through a small case study based on the trainees' organisation in order to maximise the benefit of the training. A separate course of statistical analysis may also be necessary. Such courses are often run by local University and college establishments. Training courses are also offered by colleges in project management and communication skills. A useful approach is to work with an experienced simulation consultant on early projects in order to ensure that priorities are correctly assigned to the stages of the simulation study. A common mistake is to spend too long on the model building stage before adequate consultation has been made which would achieve a fuller understanding of the problem situation.

The skills needed to successfully undertake a simulation study are varied and it has been found that the main obstacle to performing simulation in-house is not cost or training but the lack of personnel with the required technical background (Cochran et al., 1995). This need for technical skills has meant that most simulation project leaders are systems analysts, in-house simulation developers or external consultants rather than people who are closer to the process such as a shop-floor supervisor. However the need for process owners to be involved in the simulation study can be important in ensuring on-going use of the technique and that the results of the study are implemented (Greasley, 1996).

Summary

This chapter has outlined the main steps in providing a capability within the organisation to undertake simulation studies. These steps provide an assessment of the potential costs and benefits of the method and indicate the technical and organisational resources required to implement the technique.

References

Banks, J. and Gibson, R. (1997), Simulation Modelling: Some Programming Required, *IIE Solutions*, February, 26–31.

Cochran, J.K., Mackulak, G.T. and Savory, P.A. (1995), Simulation Project Characteristics in Industrial Settings, *INTERFACES*, 25: 4, 104–113.

Davies, R. and O'Keefe, R. (1989), *Simulation Modelling with Pascal*, Prentice-Hall Inc., New Jersey.

Greasley, A. (1996), The Use of Simulation Modelling in an Organisational Context, *Proceedings of the IASTED/ISMM International Conference on Modelling and Simulation*, April 25–27, IASTED.

Greasley, A. (1998), An Example of a Discrete-Event Simulation on a Spreadsheet, *SIMULATION*, 70: 3, 148–166.

Pidd, M. and Cassel, R.A. (2000), Taking Cues from Java: Featuring How Discrete Simulation and Java Can Light Up the Web, *IEEE Potentials*, February/March.

Rohrer, M. and Banks, J. (1998), Required Skills of a Simulation Analyst, *IIE Solutions*, May, 20–23.

Sprague, R.H. and Watson, H.J. (eds), (1993), *Decision Support Systems: Putting Theory into Practice*, Third Edition, Prentice-Hall Inc., New Jersey.

4

Steps in Building a Simulation Model

Introduction

This chapter outlines the steps required in undertaking a simulation project. In order to use simulation successfully a structured process must be followed. This chapter aims to show that simulation is more than just the purchase and use of a software package but a range of skills are required by the simulation practitioner or team. These include project management, client liaison, statistical skills, modelling skills and the ability to understand and map out organisational processes.

1. Formulate the Simulation Project Proposal

Simulation modelling is a flexible tool and is capable of analysing most aspects of an organisation. Therefore to ensure the maximum value is gained from using the technique it is necessary to define the areas of the organisation that are keys to overall performance and select feasible options for the technique in these areas. Another aspect to consider is the nature of the simulation model that is to be developed. In order to assist the decision-making process it is not always necessary to undertake all the stages of a simulation study. For instance the development of the process map may be used to help understanding of a problem. The level of usage of simulation is discussed in this chapter. There follows a description of project management concepts and an outline of the contents of a simulation project proposal.

Determining the Level of Usage of the Simulation Model

An important aspect in the process of building a simulation model is to recognise that there are many possible ways of modelling a system. Choices have to be made regarding the level of detail to use in modelling processes and even whether a particular process should be modelled at all. The way to make these choices is to recognise that before the model is built the objectives of the study must be defined clearly. It may even be preferable to build different versions of the model to answer different questions about the system, rather than build a single 'flexible' model that attempts to provide multiple perspectives on a problem. This is because two relatively simple models will be easier to validate and thus there will be a higher level of confidence in their results than a single complex model.

The objective of the simulation technique is to aid decision making by providing a forum for problem definition and providing information on which decisions can be made. Thus a simulation project does not necessarily require a completed computer model to be a success. At an early stage in the project proposal process the analyst and other interested parties must decide the role of the model building process within the decision-making process itself. Thus in certain circumstances the building of a computer model may not be necessary. However for many complex, interacting systems (i.e. most business systems) the model will be able to provide useful information (not only in the form of performance measures, but indications of cause and effect linkages between variables) which will aid the decision making process. The focus of the simulation project implementation will be dependent on the intended usage of the model as a decision making tool (Table 4.1).

Table 4.1. Levels of usage of a simulation model

	Level of Usage			
	Problem Definition	Demo.	Scenarios	On-going Decision Support
Level of Development	Process Map	Animation	Experimentation	Decision Support System
Level of Interaction	None	None Simple Menu	Menu	Extended Menu
Level of Integration	None	Stand-alone	Stand-alone Database	Stand-alone Database Real-Time Data

The level of usage categories are defined as follows:

Problem Definition
One of the reasons for using the simulation method is that its approach provides a detailed and systematic way of analysing a problem in order to provide information on which a decision can be made. It is often the case that ambiguities and inconsistencies are apparent in the understanding of a problem during the project proposal formulation stage. It may be that the process of defining the problem may provide the decision makers with sufficient information on which a decision can be made. In this case model building and quantitative analysis of output from the simulation model may not be required. The outcome from this approach will be a definition of the problem and possibly a process map of the system.

Demonstration
Although the decision makers may have an understanding of system behaviour, it may be that they wish to demonstrate that understanding to other interested parties. This could be to internal personnel for training purposes or to external personnel to demonstrate capability to perform to an agreed specification. The development of an animated model provides a powerful tool in communicating the behaviour of a complex system over time.

Scenarios
The next level of usage involves the development of a model and experimentation in order to assess system behaviour over a number of scenarios. The model is used to solve a number of pre-defined problems but is not intended for future use. For this reason a simple menu system allowing change of key variables is appropriate. The simulation may use internal data files or limited use of external databases.

On-going Decision Support
The most fully developed simulation model must be capable of providing decision support for a number of problems over time. This requires that the model be adapted to provide assistance to new scenarios as they arise. The menu system will need to provide the ability to change a wider range of variables for on-going use. The level of data integration may require links to company databases to ensure the model is using the latest version of data over time. Links may also be required to real-time data systems to provide on-going information on process performance. Animation facilities should be developed to assist in understanding cause and effect relationships and the effect of random events such as machine breakdowns.

Integration with shop-floor machine controllers may be necessary for real-time simulation systems. The technical hardware and software capability issues relevant to an integrated system need to be addressed at the project proposal stage to ensure a successful implementation.

If it is envisaged that the client will perform modifications to the simulation model after delivery then the issue of model re-use should be addressed. Re-use issues include ensuring detailed model code documentation is supplied and detailed operating procedures are provided. Training may also be required in model development and statistical methods. Another reason for developing a model with on-going decision-support capabilities is to increase model confidence and acceptance particularly among non-simulation experts (Muller, 1996).

Managing the Simulation Project

An important aspect of the project management process is identifying and gaining the support of personnel who have an interest in the modelling process. As stated in Chapter 3 the developer(s) in addition to the technical skills required to build and analyse the results from a model must be able to communicate effectively with people in the client organisation in order to collect relevant data and communicate model results. Roles within the project team include the following:

- *Client* – sponsor of the simulation project – usually a manager who can authorise the time and expenditure required.
- *Model User* – Person who is expected to use the model after completion by the modeller. The role of the model user will depend on the planned level of usage of the model. A model user will not exist for a problem definition exercise, but will require extended contact with the developer if the model is to be used for on-going decision-support to ensure all options (e.g. menu option facilities) have been incorporated into the design before handover.
- *Data Provider* – Often the main contact for information regarding the model may not be directly involved in the modelling outcomes. The client must ensure that the data provider feels fully engaged with the project and is allocated time for liaison and data collection tasks. In addition the modeller must be sensitive to using the data providers time as productively as possible.

The project report should contain the simulation study objectives and a detailed description of how each stage in the simulation modelling process will be undertaken. This requires a definition of both the methods to be used and any resource requirements for the project. It is important to take a structured approach to the management of the project as there are many reasons why a project could fail. These include:

- The simulation model does not achieve the objectives stated in the project plan through a faulty model design or coding.
- Failure to collect sufficient and relevant data means that the simulation results are not valid.
- The system coding or user interfaces do not permit the flexible use of the model to explore scenarios defined in the project plan.
- The information provided by the simulation does not meet the needs of the relevant decision-makers.

These diverse problems can derive from a lack of communication leading to failure to meet business needs to technical failures, such as a lack of knowledge of statistical issues in experimentation, leading to invalid model results. For this reason the simulation project manager must have an understanding of both the business and technical issues of the project. The project management process can be classified into the four areas of Estimation, Scheduling/Planning, Monitoring and Control, and Documentation.

Estimation
This entails breaking down the project into the main simulation project stages (data collection, modelling input data etc.) and allocating resources to each stage. The time required and skill type of people required along with the requirement for access to resources such as simulation software. These estimates will allow a comparison between project needs and project resources available. If there are

insufficient resources available to undertake the project then a decision must be made regarding the nature of the constraints on the project. A Resource Constrained project is limited by resources (i.e. people/software) availability. A Time Constrained project is limited by the project deadline. If the project deadline is immovable then additional resources will need to be requested in the form of additional personnel (internal or external), overtime or additional software licenses. If the deadline can be changed then additional resources may not be required as a smaller project team may undertake the project over a longer time period.

Once a feasible plan has been determined a more detailed plan of when activities should occur can be developed. The plan should take into account the difference between effort time (how long someone would normally be expected to take to complete a task) and elapse time which takes into account availability (actual time allocated to project and the number of people undertaking the task) and work rate (skill level) of people involved. In addition a time and cost specification should be presented for the main simulation project stages. A timescale for the presentation of an interim report may also be specified for a larger project. Costings should include the cost of the analyst's time and software/hardware costs. If the organisation has access to the appropriate hardware then there is the choice between a 'run-time' license (if available) providing use of the model but not the ability to develop new models. A full license is appropriate if the organisation wishes to undertake development work in-house. Although an accurate estimate of the timescale for project completion is required the analyst or simulation client needs to be aware of several factors that may delay the project completion date. The most important factor in the success of a simulation project is to ensure that appropriate members of the organisation are involved in the simulation development. The simulation provides information on which decisions are made within an organisational context, so involvement is necessary of interested parties to ensure confidence and implementation of model results. The need for clear objectives is essential to ensure the correct systems components are modelled at a suitable level of detail. Information must also be supplied for the model build from appropriate personnel to ensure important detail is not missing and false assumptions regarding model behaviour are not made. It is likely that during the simulation process, problems with the system design become apparent that require additional modelling and/or analysis. Both analyst and client need to separate between work for the original design and additional activity. The project specification should cover the number of experimental runs that are envisaged for the analysis. Often the client may require additional scenarios tested, which again should be agreed at a required additional time/cost.

Scheduling/Planning
Scheduling involves determining when activities should occur. Steps given in the simulation study are sequential, but in reality they will overlap – the next stage starts before the last one is finished – and are iterative e.g. validate part of the model, go back and collect more data, model build, validate again. This iterative process of building more detail into the model gradually is the recommended approach but can make judging project progress difficult.

Monitoring and Control
A network plan is useful for scheduling overall project progress and ensuring on-time completion but the reality of iterative development may make it difficult to judge actual progress.

Documentation
Interim progress reports are issued to ensure the project is meeting time and cost targets. Documents may also be needed to record any changes to the specification agreed by the project team. Documentation provides traceability. For example data collection sources and content should be available for inspection by users in future in order to ensure validation. Documentation is also needed of all aspects of the model such as coding and the results of the simulation analysis.

The Simulation Project Proposal

The requirements for each section of the simulation project proposal are now given.

Study Objectives
A number of specific study objectives should be derived which will provide a guide to the data needs of the model, set the boundaries of the study (scope), the level of modelling detail and define the experimentation analysis required. It is necessary to refine the study objectives until specific scenarios defined by input variables and measures that can be defined by output variables can be specified. General improvement areas for a project include aspects such as the following:

- Changes in Process Design – Changes to routing, decision points and layout.
- Changes in Resource Availability – Shift patterns, equipment failure.
- Changes in Demand – Forecast pattern of demand on the process.

Many projects will study a combination of the above, but it is important to study each area in turn to establish potential subjects for investigation at the project proposal stage. The next step is to define more specifically the objectives of the study. For example a manufacturing facility may wish to study the effect of machine breakdown on the output of a production line. The objective may be written thus:

> 'The simulation model will provide a sensitivity analysis of the breakdown rate of machine x on the output units of production line y.'

From this the simulation analyst can derive the following requirements from the simulation model specification. Important information on model detail (machine availability), input requirements (breakdown behaviour of machine) and performance measures required (output units of production line) can be implied from the objective stated. The project proposal should contain a number of objectives at such a level of detail which allows the simulation analyst to derive an outline model specification upon which a quotation in terms of time and cost can be prepared. For instance the previous statement requires more clarification on the exact nature of the analysis of breakdowns. For example will an analysis using a simple graph showing breakdown rate vs. output suffice? Is it required to show breakdown behaviour under a number of scenarios e.g. additional preventative maintenance measures? These factors may require simply additional simulation runs, or major changes to the model design – so an iterative process of re-defining model objectives between analyst and user is required at this stage.

Once the objectives and experiments have been defined the scope and level of detail can be ascertained. The model scope is the definition of the boundary between what is to be included in the model and what is considered external to the specification. Once the scope has been determined it is necessary to determine the level of detail in which to model elements within the model scope. In order to keep the model complexity low only the minimum model scope and level of detail should be used. Regarding model scope there can be a tendency for the model user to want to include every aspect of a process. However this may entail building such a complex model that the build time and the complexity of interpreting model results may lead to a failed study. Regarding model detail, judgement is required in deciding what elements of the system should be eliminated or simplified to minimise unnecessary detail. An iterative process of model validation and addition of model detail should be followed. Strategies for minimising model detail include:

- Modelling a group of processes by a single process – often the study requires no knowledge of the internal mechanisms within a process and only the process time delay is relevant to overall performance.
- Assuming continuous resource availability – the modelling of shift patterns of personnel or maintenance patterns for machinery may not always be necessary if their effect on performance is small.
- Infrequent events such as personnel absence through sickness or machine breakdown may occur so infrequently that they are not necessary to model.

What is important is that any major assumptions made by the developer at the chosen level of detail are stated explicitly in the simulation report, so that the user is aware of them.

Data Collection and Process Mapping
Once the simulation project objectives have been defined, and the scope and level of detail set, the modeller should prepare a specification of the data required for the model build. It is useful at this stage to identify the source of the information, its form (e.g. documentation, observation and interview) and any personnel responsible for supplying the relevant information. A process map specification

should define what processes will be mapped. The process map should provide a medium for obtaining information from a variety of viewpoints regarding the system being organised. In particular, issues of system boundaries (i.e. what to include and what to omit from the analysis) can be addressed (see Chapter 4 for more details).

Modelling Input Data
A specification of the type of statistical analysis used for modelling input variables and process durations should be made. A trace driven model will require no statistical analysis of input data, but a forecasting model may require that the data is fitted to a suitable probability distribution. The level of data analysis will depend on the study objectives, time constraints, and the amount of raw data available (see Chapter 5 for more details).

Building the Model
The simulation software type or the software package used to construct the model should be specified. To allow ongoing use the software should either be available within the organisation or a run-time facility provided (see Chapter 6 for more details).

Validation and Verification
Verification or debugging time can be difficult to predict but some estimation of verification time can be made from the estimated complexity and size of the proposed model. Validation will require the analyst to spend time with people who are familiar with the model being studied to ensure model behaviour is appropriate. A number of meetings may be necessary to ensure that the simulation model has sufficient credibility with potential users. Sensitivity analysis may be required to validate a model of a system that does not currently exist. The type of sensitivity analysis envisaged should be defined in the project proposal (see Chapter 7 for more details).

Experimentation and Analysis
Experimentation and analysis aims to study the effects of changes in input variables (i.e. scenarios defined in the objectives) have on output variables (i.e. performance measures defined in the objectives) in the model. The number of experiments should be clearly defined as each experiment may take a substantial amount of analysis time. For each experiment the statistical analysis required should be defined. This could include aspects such as statistical tests, use of common random numbers and graphical analysis. The use of a terminating or steady-state analysis should be stated. For example an analysis of breakdown behaviour may require the following experimental analysis.

> 'The simulation will be run 20 times at 5 settings of the breakdown rate and the process output noted at each setting. The results will be presented in tabular and graphical format.'

Implementation
The results of the simulation study must be presented in report form (Chapter 9), which should include full model documentation, study results and recommendations for further studies. An implementation plan may also be specified. The report can be supplemented by a presentation to interested parties. The duration and cost of both of these activities should be estimated. Further allocation of time and money may be required for aspects such as user training, run time software license and telephone support from an external consultant (see Chapter 9 for more details).

2. Data Collection

The collection of data is one of the most important and challenging aspects of the simulation modelling process. A model which accurately represents a process will not provide accurate output data unless the input data has been collected and analysed in an appropriate manner. Data requirements for the model can be grouped into two areas. In order to construct the process map which describes the logic of the model (i.e. how the process elements are connected) the process routing and decision points are required as follows:

Logic Data Required for the Process Map

- *Process Routing* – All possible routes of people/components/data through the system.
- *Decision Points* – Decision points can be modelled by conditional (if.... then x, else y) or probability (with 0.1, x; with 0.5, y; else z) methods.

In order to undertake the model building stage further data is required in terms of the process durations, resource availability schedules, demand patterns and the process layout.

Additional Data Required for the Simulation Model

- *Process Timing* – Durations for all relevant processes (e.g. customer service time at a bank – but not the queuing time). Can be a data sample from which a probability distribution is derived.
- *Resource Availability* – Resource availability schedules for all relevant resources, including effects of shift patterns and breakdowns etc.
- *Demand Pattern* – A schedule of demand which 'drives' the model (e.g. customer arrivals)
- *Process Layout* – Diagram/schematic of the process which can be used to develop the simulation animation display.

Be sure to distinguish between input data which is what should be collected and output data which is dependent on the input data values. For example customer arrival times would usually be input data while customer queue time is output data, dependent on input values such as customer arrival rate. However, although we would not enter the data collected on queue times into our model we could compare these times to the model results to validate the model.

The required data may not be available in a suitable format, in which case the analyst must either collect the data or find a way of working around the problem. In order to amass the data required it is necessary to use a variety of data sources shown in the Table 4.2.

Table 4.2. Sources of Data

Data Source	Example
Historical Records	diagrams, schematics, schedules
Observations	time studies, walkthroughs
Interviews	discussion of process steps
Process Owner/Vendor Estimates	process time estimates

Historical Records
A mass of data may be available within the organisation regarding the system to be modelled in the form of schematic diagrams, production schedules, shift patterns etc. This data may be in a variety of formats including paper and electronic (e.g. held on a database). However this data may not be in the right format, be incomplete or not relevant for the study in progress. The statistical validity of the data may also be in doubt.

Observations
A walkthrough of the process by the analyst is an excellent way of gaining an understanding of the process flow. Time studies can also be used to estimate process parameters when current data is not available.

Interviews
An interview with the process owner can assist in the analysis of system behaviour which may not always be documented

Process Owner/Vendor Estimate
Process owner and vendor estimates are used most often when the system to be modelled does not exist and thus no historical data or observation is possible. This approach has the disadvantage of relying on the ability of the process owner (e.g. machine operator, clerk) in remembering past performance. If possible a questionnaire can be used to gather estimates from a number of process owners and the data statistically analysed. Vendor information may also be based on unrealistic assumptions of ideal conditions for equipment operation. If no

estimates can be made then the objectives relating to those aspects may need to be changed to remove that aspect of the analysis from the project.

As with other stages of a simulation project data collection is an iterative process with further data collected as the project progresses. For instance statistical tests during the modelling of input data or experimentation phases of development may suggest a need to collect further data in order to improve the accuracy of results. Also the validation process may expose inaccuracies in the model which require further data collection activities. Thus it should be expected that data collection activities will be on-going throughout the project as the model is refined.

3. Process Mapping

A process map (also called a conceptual model) should be formulated in line with the scope and level of detail defined within the project specification (Chapter 3). An essential component of this activity is to construct a diagrammatic representation of the process in order to provide a basis for understanding between the simulation developer and process owner. Two diagramming methods used in discrete-event simulation are activity cycle diagrams and process maps. Activity cycle diagrams can be used to represent any form of simulation system. Process maps are most suited to representing a process-interaction view that follows the life cycle of an entity (e.g. customer, product) through a system comprising a number of activities with queuing at each process (e.g. waiting for service, equipment). Most simulation applications are of this type and the clear form of the process map makes it the most suitable method in these instances.

Two main problems associated with data are that little useful data is available (when modelling a system that does not yet exist for example) or that the data is not in the correct format. If no data exist you are reliant on estimates from vendors or other parties, rather than samples of actual performance, so this needs to be emphasized during the presentation of any results. An example of data in the wrong format is a customer service time calculated from entering the service queue to completion of service. This data could not be used to approximate the customer service time in the simulation model as you require the service time only. The queuing time will be generated by the model as a consequence of the arrival rate and service time parameters. In this case the client may assume that your data requirements have been met and will specify the time and cost of the simulation project around that. Thus it is important to establish as soon as possible the actual format of the data and its suitability for your needs to avoid misunderstandings later.

A number of factors will impact on how the data collection process is undertaken including the time and cost within which project must be conducted. Compromises will have to be made on the scale of the data collection activity and so it is important to focus effort on areas where accuracy is important for simulation results and to make clear assumptions made when reporting simulation results. If it has not been possible to collect detailed data in certain areas of the process, it is not sensible to then model in detail that area. Thus there is a close relationship between simulation objectives, model detail and data collection needs. If the impact of the level of data collection on results is not clear, then it is possible to use sensitivity analysis (i.e. trying different data values) to ascertain how much model results are affected by the data accuracy. It may be then necessary to either undertake further data collection or quote results over a wide range.

Activity Cycle Diagrams

Activity Cycle diagrams can be used to construct a conceptual model of a simulation which uses the event, activity or process orientation. The diagram aims to show the life-cycles of the components in the system. Each component is shown in either of two states, the *dead* state is represented by a circle and the *active* state is represented by a rectangle. Each component can be shown moving through a number of dead and active states in a sequence that must form a loop. The dead state relates to a conditional ('C') event where the component is waiting for something to happen such as the commencement of a service for example. The active state relates to a bound ('B') event or a service process for example. The duration of the active state is thus known in advance whilst the duration of the dead state cannot be known, because it is dependent on the behaviour of the whole system.

Process Maps

The construction of a process flow diagram is a useful way of understanding and documenting any business process and showing the interrelationships between activities in a process. These diagrams have become widely used in Business Process Reengineering (BPR) projects and the use of process mapping in this context is evaluated in Peppard and Rowland (1995). For larger projects it may be necessary to represent a given process at several levels of detail. Thus a single activity may be shown as a series of sub-activities on a separate diagram. In simulation projects this diagram is often referred to as the simulation conceptual model and the method is particularly suitable when using process oriented simulation languages and Visual Interactive Modelling systems.

4. Modelling Input Data

It is important to model randomness in such areas as arrival times and process durations. Taking an average value will not give the same behaviour. Queues are often a function of the *variability* of arrival and process times and not simply a consequence of the relationship between arrival interval and process time. The method of modelling randomness used in the simulation will be dependent on the amount of data collected on a particular item. For less then 20 data points a mean value or theoretical distribution must be estimated. Larger samples allow the user to fit the data to a theoretical distribution or to construct an empirical distribution.

Theoretical and empirical distributions are classified as either continuous or discrete. Continuous distributions can return any real value quantity and are used to model arrival times and process durations. Discrete distributions return only whole number or integer values and are used to model decision choices or batch sizes. Table 4.3 provides guidance on possible methods for modelling randomness.

Table 4.3. Modelling Methods by Number of Data Points

Data Points	Suggested Modelling Method
Less than 20	Could use mean, exponential, triangular, normal or uniform
20+	Fit theoretical distribution
200+	Construct empirical distribution
Historical	Trace

Less than 20 Data Points: Estimation

If it is proposed to build a model of a system that has not been built or there is no time for data collection then an estimate must be made. This can be achieved by questioning interested parties such as the process owner or the equipment vendor for example. A sample size of below 20 is probably too small to fit a theoretical distribution with any statistical confidence although it may be appropriate to construct a histogram to assist in finding a representative distribution.

The simplest approach is to simply use a fixed value to represent the data representing an estimate of the mean. Otherwise a theoretical distribution may be chosen based on knowledge and statistical theory. Statistical theory suggests that if the mean value is not very large, interarrival times can be simulated using the exponential distribution. Service times can be simulated using a uniform or symmetric triangular distribution with the minimum and maximum values at percentage variability from the mean. For example a mean of 100 with a variability of +/− 20% would give values for a triangular distribution of 80 for minimum, 100 for mode and 120 for maximum. The normal distribution may be used when an unbounded (i.e. the lower and upper levels are not specified) shape is required. The normal distribution requires mean and standard deviation parameters. When only the minimum and maximum values are known and behaviour between those values is not known a uniform distribution generates all values with an equal likelihood.

20+ Data Points: Deriving a Theoretical Distribution

For 20+ data points a theoretical distribution can be derived. The standard procedure to match a sample distribution to a theoretical distribution is to construct a histogram of the data and compare the shape of the histogram with a range of theoretical distributions. Once a potential candidate is found it is necessary to estimate the parameters of the distribution which provides the closest fit. The relative 'goodness of fit' can be determined by using an appropriate statistical method.

200+ Data Points: Constructing an Empirical Distribution

For more than 200 data points the option of constructing a user-defined distribution is available. An empirical or user-defined distribution is a distribution that has been obtained directly from the sample data. An empirical distribution is usually chosen if a reasonable fit cannot be made with the data and a theoretical distribution. It is usually necessary to have in excess of 200 data points to form an empirical distribution. In order to convert the sample data into an empirical distribution the data is converted into a cumulative probability distribution using the following steps.

i. Sort values into ascending order
ii. Group identical values (discrete) or group into classes (continuous)
iii. Compute the relative frequency of each class
iv. Compute the cumulative probability distribution of each class

Historical Data Points

A simulation driven by historical data is termed a 'trace-driven' simulation. An example would be using actual arrival times of customers in a bank directly in the simulation model. The major drawback of this approach is that it prevents the simulation from being used in the 'what-if' mode as only the historical data is modelled. It also does not take account of the fact that in the future the system will most likely encounter conditions out of the range of the sample data used. This approach can be useful however in validating model performance when the behaviour of the model can be compared to the real system with identical data.

5. Building the Model

This involves using computer software to translate the process map into a computer simulation model which can be 'run' to generate model results. This will entail the use of simulation software such as ARENA, WITNESS or SIMUL8. Kelton et al. (2007) covers the use of ARENA and Greasley (2004) covers all three of the above software packages.

6. Validation and Verification

Before experimental analysis of the simulation model can begin it is necessary to ensure that the model constructed provides a valid representation of the system we are studying. This process consists of verification and validation of the simulation model. Verification refers to ensuring that the computer model built using the simulation software is a correct representation of the process map of the system under investigation. Validation concerns ensuring that the assumptions made in the process map about the real-world system are acceptable in the context of the simulation study. Both topics will now be discussed in more detail.

Verification

Verification is analogous to the practice of 'debugging' a computer program. Thus many of the following techniques will be familiar to programmers of general-purpose computer languages.

Model Design

The task of verification is likely to become greater with an increase in model size. This is because a large complex program is both more likely to contain errors and these errors are less likely to be found. Due to this behaviour most practitioners advise on an approach of building a small simple model, ensuring that this works correctly, and then gradually adding embellishments over time. This approach is intended to help limit the area of search for errors at any one time. It is also important to ensure that unnecessary complexity is not incorporated in the model design. The design should incorporate only enough detail to ensure the study objectives and not attempt to be an exact replica of the real-life system.

Structured Walkthrough

This enables the modeller to incorporate the perspective of someone outside the immediate task of model construction. The walkthrough procedure involves talking through the program code with another individual or team. The process may bring fresh insight from others, but the act of explaining the coding can also help the person who has developed the code to discover their own errors. In discrete-event simulation code is executed non-sequentially and different coding blocks are executing simultaneously. This means that the walkthrough may best be conducted by following the 'life-history' of an entity through the simulation coding, rather than a sequential examination of coding blocks.

Test Runs

Test runs of a simulation model can be made during program development to check model behaviour. This is a useful way of checking model behaviour as a defective model will usually report results (e.g. machine utilisation, customer wait times) which do not conform to expectations, either based on the real system performance or common-sense deductions. It may be necessary to add performance measures to the model (e.g. costs) for verification purposed, even though they may not be required for reporting purposes. One approach is to use historical (fixed) data, so model behaviour can be isolated from behaviour caused by the use of random varieties in the model. It is also important to test model behaviour under a number of scenarios, particularly boundary conditions that are likely to uncover erratic behaviour. Boundary conditions could include minimum and maximum arrival rates, minimum and maximum service times and minimum and maximum rate of infrequent events (e.g. machine breakdowns).

Trace Analysis

Due to the nature of discrete-event simulation it may be difficult to locate the source of a coding error. Most simulation packages incorporate an entity trace facility that is useful in providing a detailed record of the life-history of a particular entity. The trace facility can show the events occurring for a particular entity or all events occurring during a particular time-frame. The trace analysis facility can produce a large amount of output so it is most often used for detailed verification.

The animation facilities of simulation software packages provide a powerful tool in aiding understanding of model behaviour. The animation enables the model developer to see many of the model components and their behaviour simultaneously. A 'rough-cut' animated drawing should be sufficient at the testing stage for verification purposes. To aid understanding model components can be animated which may not appear in the final layout presented to a client. The usefulness of the animation technique will be maximised if the animation software facilities permit reliable and quick production of the animation effects.

It is important to document al elements in the simulation to aid verification by other personnel or at a later date. Any general–purpose or simulation coding should have comments attached to each line of code. Each object within a model produced on a Visual Interactive Modelling system requires comments regarding its purpose and details of parameters and other elements.

Validation

A verified model is a model which operates as intended by the modeller. However this does not necessarily mean that it is a satisfactory representation of the real system for the purposes of the study. Validation is about ensuring that model behaviour is close enough to the real-world system for the purposes of the simulation study. Unlike verification, the question of validation is one of judgement. Ideally the model should provide enough accuracy for the measures required whilst limiting the amount of effort required achieving this. For most systems of any complexity this aim can be achieved in a number of ways and a key skill of the simulation developer is finding the most efficient way of achieving this goal. Pegden (1995) outlines three aspects of validation:

- Conceptual Validation – Does the model adequately represent the real-world system?
- Operational Validity – Are the model generated behavioural data characteristic of the real-world system behavioural data?
- Believability – Does the simulation model's ultimate user have confidence in the model's results

Conceptual Validity

Conceptual validation involves ensuring that the model structure and elements are correctly chosen and configured in order to adequately represent the real-world system. As we know that the model is a simplification of the real-world then there is a need for a consensus around the form of the conceptual model between the model builder and the user. To ensure a credible model is produced the model builder should discuss and obtain information from people familiar with the real-world system including operating personnel, industrial engineers, management, vendors and documentation. They should also observe system behaviour over time and compare with model behaviour and communicate with project sponsors throughout the model build to increase credibility.

Operational Validity

This involves ensuring that the results obtained from the model are consistent with real-world performance. A common way of ensuring operational validity is to use the technique of sensitivity analysis to test the behaviour of the model under various scenarios and compare results with real-world behaviour. The technique of common random numbers can be used to isolate changes due to random variation. The techniques of experimental design can also be employed to conduct sensitivity analysis over two or more factors. Note that for validation purposes these tests are comparing simulation performance with real-world performance while in the context of experimentation they are used to compare simulation behaviour under different scenarios.

Sensitivity analysis can be used to validate a model but it is particularly appropriate if a model has been built of a system which does not exist as the data has been estimated and cannot be validated against a real system. In this case the main task is to determine the effect of variation in this data on model results. If there is little variation in output as a consequence of a change in input then we can be reasonably confident in the results. It should also be noted that an option may be to conduct sensitivity analysis on sub-systems of the overall system being modelled which do exist. This emphasises the point that the model should be robust enough to provide a prediction of what would happen in the real system under a range of possible input data. The construction and validation of the model should be for a particular range of input values defined in the simulation project objectives. If the simulation is then used outside of this predefined range the model must be re-validated to ensure additional aspects of the real system are incorporated to ensure valid results.

An alternative to comparing the output of the simulation to a real system output is to use actual historical data in the model, rather than derive a probability distribution. Data collected could be used for elements such as customer arrival times and service delays. By comparing output measures across identical time periods it should be possible to validate the model. Thus the structure or flow of the model could be validated and then probability distributions entered for random elements. Thus any error in system performance could be identified as either a logic error or from an inaccurate distribution. The disadvantage of this method is that for a model of any size the amount of historical data needed will be substantial. It is also necessary to read this data, either from a file or array, requiring additional coding effort.

Sensitivity analysis should be undertaken by observing the output measure of interest with data set to levels above and below the initial set level for the data. A graph may be used to show model results for a range of data values if detailed analysis is required. (e.g. a non-linear relationship between a data value and output measure is apparent). If the model output does not show a significant change in value in response to the sensitivity analysis then we can judge that the accuracy of the estimated value will not have a significant effect on the result.

If the model output is sensitive to the data value then preferably we would want to increase the accuracy of the data value estimate. This may be undertaken by further interviews or data collection. In any event the simulation analysis will need to show the effect of model output on a range of data values. Thus for an estimated value we can observe the likely behaviour of the system over a range of data values within which the true value should be located. Further sensitivity analysis may be required on each of these values to separate changes in output values from random variation.

When it is found that more than one data value has an effect on an output measure, then the effects of the individual and combined data values should be assessed. This will require 3^k replications to measure the minimum, initial and maximum values for k variables. The use of fractional factorial designs techniques (Law and Kelton, 2000) may be used to reduce the number of replications required.

Believability

In order to ensure implementation of actions recommended as a result of simulation experiments requires that the model output is seen as credible from the simulation user's point of view. This credibility will be enhanced by close co-operation between model user and client throughout the simulation study. This involves agreeing clear project objectives explaining the capabilities of the technique to the client and agreeing assumptions made in the process map. Regular meetings of interested parties, using the simulation animation display to provide a discussion forum, can increase confidence in model results. Believability emphasises how there is no one answer to achieving model validity and the perspective of both users and developers need to be satisfied that a model is valid.

7. Experimentation and Analysis

The stochastic nature of simulation means that when a simulation is run the performance measures generated are a sample from a random distribution. Thus each simulation 'run' will generate a different result, derived from the randomness which has been modelled. In order to interpret the results (i.e. separate the random changes in output from changes in performance) statistical procedures are outlined in this chapter which are used for the analysis of the results of runs of a simulation. There are two types of simulation system that need to be defined, each requiring different methods of data analysis.

Terminating systems run between pre-defined states or times where the end state matches the initial state of the simulation (i.e. a simulation of a shop from opening to closing time). Non-terminating systems do not reach pre-defined states or times. In particular the initial state is not returned to, for example a manufacturing facility. Most service organisations tend to be terminating systems which close at the end of each day with no in-process inventory (i.e. people waiting for service) and thus return to the 'empty' or 'idle' state they had at the start of that day. Most manufacturing organisations are non-terminating with inventory in the system that is awaiting a process. Thus even if the facility shuts down temporarily it will start again in a different state to the previous start-state (i.e. the inventory levels define different starting conditions). However the same system may be classified as terminating or non-terminating depending on the objectives of the study. Before a non-terminating system is analysed the bias introduced by the non-representative starting conditions must be eliminated to obtain what are termed steady-state conditions from which a representative statistical analysis can be undertaken.

Statistical Analysis for Terminating Systems

This section will provide statistical tools to analyse either terminating systems or the steady-state phase of non-terminating systems. The statistics relevant to both the analysis of a single model and comparing between different models will now be outlined in turn.

The output measure of a simulation model is a random variable and so we need to conduct multiple runs (replications) of the model to provide us with a sample of its value. When a number of replications have been undertaken the sample mean can be calculated by averaging the measure of interest (e.g. time in queue) over the number of replications made. Each replication will use of different set of random numbers and so this procedure is called the method of independent replications.

Establishing a Confidence Interval

To assess the precision of our results we can compute a confidence interval or range around the sample mean that will include, to a certain level of confidence, the true mean value of the variable we are measuring. Thus confidence intervals provide a point estimate of the expected average (average over infinite number of replications) and an idea of how precise this estimate is. The confidence interval will fall as replications increase. Thus a confidence interval does not mean that say 95% of values fall within this interval, but that we are 95% sure that the interval contains the expected average.

For large samples (replications) of over around 50 the normal distribution can be used for the computations. However the sample size for a simulation experiment will normally be less than this with 10 replications of a simulation being common. In this case provided the population is approximately normally distributed the sampling distribution follows a t-distribution.

Both the confidence interval analysis and the t-tests presented later for comparison analysis assume the data measured is normally distributed. This assumption is usually acceptable if measuring an average value for each replication as the output variable is made from many measurements and the central limit theorem applies. However the central limit theorem applies for a large sample size and the definition of what constitutes a large sample depends partly on how close the actual distribution of the output variable is to the normal distribution. A histogram can be used to observe how close the actual distribution is to the normal distribution curve.

Comparing Alternatives
When comparing between alternative configurations of a simulation model we need to test whether differences in output measures are statistically significant or if differences could be within the bounds of random variation. Alternative configurations which require this analysis includes:

- changing input parameters (e.g. changing arrival rate)
- changing system rules (e.g. changing priority at a decision point)
- changing system configuration (comparing manual vs. automated system)

Whatever the scale of the differences between alternative configurations there is a need to undertake statistical tests. The tests will be considered for comparing between two alternatives and then between more than two alternatives.

The following assumptions are made when undertaking the tests:

- The data collected *within* a given alternative are independent observations of a random variable. This can be obtained by each replication using a different set of random numbers (Method of Independent Replications)

- The data collected *between* alternatives are independent observations of a random variable. This can be obtained by using a separate number stream for each alternative. This can be implemented by changing the seeds of the random number generator between runs. Note however that certain tests use the ability to use common random numbers for each simulation run in their analysis (see paired t-test using common random numbers).

Hypothesis Testing
When comparing simulation scenarios we want to know if the results of the simulation for each scenario are different because of random variability or because of an actual change in performance. In statistical terms we can do this using a hypothesis test to see if the sample means of each scenario differ.

A hypothesis test makes an assumption or hypothesis (termed the null hypothesis, H_0) and tries to disprove it. Acceptance of the null hypothesis implies that there is insufficient evidence to reject it (it does not prove that it is true).

Rejection of the null hypothesis however means that the alternative hypothesis (H_1) is accepted. The null hypothesis is tested using a test statistic (based on an appropriate sampling distribution) at a particular significance level α which relates to the area called the critical region in the tail of the distribution being used. If the test statistic (which we calculate), lies in the critical region, the result is unlikely to have occurred by chance and so the null hypothesis would be rejected. The boundaries of the critical region, called the critical values, depend on whether the test is two-tailed (we have no reason to believe that a rejection of the null hypothesis implies that the test statistic is either greater or less than some assumed value) or one-tailed (we have reason to believe that a rejection of the null hypothesis implies that the test statistic is either greater or less than some assumed value).

We must also consider the fact that the decision to reject or not reject the null hypothesis is based on a probability. Thus at a 5% significance level there is a 5% chance that H_0 will be rejected when it is in fact true. In statistical terminology this is called a type I error. The converse of this is accepting the null hypothesis when it is in fact false, called a type II error. Usually α values of 0.05 (5%) or 0.01 (1%) are used. An alternative to testing at a particular significance level is to calculate the p-value which is the lowest level of significance at which the observed value of the test statistic is significant. Thus a p-value of 0.045 (indicating a type I error occurring 45 times out of 1000) would show that the null hypothesis would be rejected at 0.05, but only by a small amount.

Paired t-test
The test calculates the difference between the two alternatives for each replication. It tests the hypothesis that if the data from both models is from the same distribution then the mean of the differences will be zero.

Paired t-test Using Common Random Numbers
The idea of using common random numbers (CRN) is to ensure that alternative configurations of a model differ only due to those configurations and not due to the different random number sets used to drive the random variables within the model. It is important that synchronisation of random variables occurs across the model configurations, so the use of a dedicated random number stream for each random variety is recommended. Again as with other variance reduction techniques the success of the method will be dependent on the model structure and there is no certainty that variance will be actually reduced. Another important point is that by driving the alternative model configuration with the same random numbers we are assuming that they will behave in a similar manner to large or small values of the random variables driving the models. In general it is advisable to conduct a pilot study to ensure that the CRN technique is in fact reducing the variance between alternatives.

Because the output from a simulation model is a random variable the variance of that output will determine the precision of the results obtained from it. Statistical techniques to reduce that variance may be used to either obtain smaller confidence intervals for a fixed amount of simulating time or achieve a desired confidence interval with a smaller amount of simulating

A variety of variance reduction techniques (VRT) are discussed in Law and Kelton (2000). The use of Common Random Numbers (CRN) in conjunction with the paired t-test is described for comparing alternative system configurations. The paired-t approach assumes the data is normally distributed (see 'testing for normality' section to check if the data is normally distributed) but does not assume all observations from the two alternatives are independent of each other, as does the two-sample-t approach.

One-way ANOVA

One-way analysis of variance (ANOVA) is used to compare the means of several alternative systems. Several replications are performed of each alternative and the test attempts to determine whether the variation in output performance is due to differences *between* the alternatives or due to inherent randomness *within* the alternatives themselves. This is undertaken by comparing the ratio of the two variations with a test statistic. The test makes the following assumptions:

- Independent data both within and between the data sets
- Observations from each alternative are drawn from a normal distribution
- The normal distributions have the same variance

The first assumption implies the collection of data using independent runs or the batch means technique, but precludes the use of variance reduction techniques (e.g. common random numbers). The second assumption implies that each output measure is the mean of a large number of observations. This assumption is usually valid but can be tested with the Chi-Square or Kologomoriv-Smirnov test if required. The third assumption may require an increase in replication run length to decrease the variances of mean performance. The F-test can be used to test this assumption if required. The test finds if a significant difference between means is apparent but does not indicate if all the means are different, or if the difference is between particular means. To identify where the differences occur then tests such as Tukeys HSD test may be used (Black, 1992). Alternatively confidence intervals between each combination can provide an indication (Law and Kelton, 2000).

Statistical Analysis for Non-terminating Systems

The previous section considered statistical analysis for terminating systems. This section provides details of techniques for analysing steady-state systems in which the start conditions of the model are not returned to. These techniques involve more complex analysis then for a terminating system and so consideration should be given to treating the model as a terminating system if at all possible.

A non-terminating system generally goes through an initial transient phase and then enters a steady-state phase when its condition is independent of the simulation starting conditions. This behaviour could relate to a manufacturing system starting from an empty ('no-inventory') state and then after a period of time moving to a stabilised behaviour pattern. A simulation analysis will be directed towards measuring performance during the steady-state phase and avoiding measurements during the initial transient phase. The following methods of achieving this are discussed:

Setting Starting Conditions

This approach involves specifying start conditions for the simulation which will provide a quick transition to steady-state conditions. Most simulations are started in an empty state for convenience but by using knowledge of steady-state conditions (e.g. stock levels) it is possible to reduce the initial bias phase substantially. The disadvantage with this approach is the effort in initialising simulation variables, of which there may be many, and when a suitable initial value may not be known. Also it is unlikely that the initial transient phase will be eliminated entirely. For these reasons the warm-up period method is often used.

Using a Warm-up Period

Instead of manually entering starting conditions this approach uses the model to initialise the system and thus provide starting conditions automatically. This approach discards all measurements collected on a performance variable before a preset time in order to ensure that no data is collected during the initial phase. The point at which data is discarded must be set late enough to ensure that the simulation has entered the steady-state phase, but no so late that insufficient data points can be collected for a reasonably precise statistical analysis. A popular method of choosing the discard point is to visually inspect the simulation output behaviour of the variable over time. Welch (1983) suggests a procedure using the moving average value in order to smooth (i.e. separate the long-term trend values from short-term fluctuations) the output response. It is important to ensure the model is inspected over a time period which allows infrequent events (e.g. machine breakdown) to occur a reasonable number of times.

In order to determine the behaviour of the system over time and in particular to identify steady-state behaviour a performance measure must be chosen. A value such as work-in-progress (WIP) provides a useful measure of overall system behaviour. In a manufacturing setting this could relate to the amount of material within the system at any one time. In a service setting (as is the case with the bank clerk model) the work-in-progress measure represents the number of customers in the system. While this measure will vary over time in a steady-state system the long-term level of WIP should remain constant.

Using an Extended Run Length

This approach simply consists of running the simulation for an extended run length, so reducing the bias introduced on output variables in the initial transient phase. This approach is best applied in combination with one or both of the previous approaches.

Batch Means Analysis

To avoid repeatedly discarding data during the initial transient phase for each run, an alternative approach allows all data to be collected during one long run. The batch means method consists of making one very long run of the simulation and collecting data at intervals during the run. Each interval between data collection is termed a batch. Each batch is treated as a separate run of the simulation for analysis. The batch means method is suited to systems that have very long warm-up periods and so avoidance of multiple replications is desirable. However with the increase in computing power available this advantage has diminished with run lengths needing to be extremely long in order to slow down analysis considerably. The batch means method also requires the use of statistical analysis methods which are beyond the scope of this book (see Law and Kelton, 2000).

8. Presentation of Results

For each simulation study the simulation model should be accompanied by a project report, outlining the project objectives and providing the results of experimentation. Discussion of results and recommendations for action should also be included. Finally a further work section will communicate to the client any possible developments and subsequent results it is felt could be obtained from the model. If there are a number of results to report, an appendix can be used to document detailed statistical work for example. This enables the main report to focus on the key business results derived from the simulation analysis. A separate technical document may also be prepared which may incorporate a model and/or model details such as key variables and a documented coding listing. Screen shots of the model display can also be used to show model features. If the client is expected to need to develop the code in-house then a detailed explanation of model coding line-by-line will be required. The report structure should contain the following elements:

- Introduction
- Description of the Problem Area
- Model Specification
- Simulation Experimentation
- Results
- Conclusions and Recommendations
- Further Studies
- Appendices: Process Logic, Data Files, Model Coding

A good way of 'closing' a simulation project is to organise a meeting of interested parties and present a summary of the project objectives and results. Project documentation can also be distributed at this point. This enables discussion of the outcomes of the project with the client and provides an opportunity to discuss further analysis. This could be in the form of further developments of the current model ('updates' or 'new phase') or a decision to prepare a specification for a new project.

9. Implementation

It is useful to both the simulation developer and client if an implementation plan is formed to undertake recommendations from the simulation study. Implementation issues will usually be handled by the client, but the simulation developer may be needed to provide further interpretation of results or conduct further experimentation. Changes in the system studied may also necessitate model modification. The level of support at this time from the developer may range from a telephone 'hotline' to further personal involvement specified in the project report. Results from a simulation project will only lead to implementation of changes if the credibility of the simulation method is assured. This is achieved by ensuring each stage of the simulation project is undertaken correctly.

Organisational Context of Implementation

A simulation modelling project can use extensive resources both in terms of time and money. Although the use of simulation in the analysis of a one-off decision, such as investment appraisal, can make these costs low in terms of making the correct decision, the benefits of simulation can often be maximised by extending the use of the model over a period of time. It is thus important that during the project proposal stage that elements are incorporated into the model and into the implementation plan that assist in enabling the model to provide on-going decision support. Aspects include:

- Ensure that simulation users are aware at the project proposal stage that the simulation is to be used for on-going decision support and will not be put to one side once the immediate objectives are met.
- Ensure technical skills are transferred from simulation analysts to simulation users. This ensures understanding of how the simulation arrives at results and its potential for further use in related applications.
- Ensure communication and knowledge transfer from simulation consultants and industrial engineers to business managers and operational personnel.

The needs of managerial and operational personnel are now discussed in more detail.

Managerial Involvement
The cost associated with a simulation project means that the decision of when and where to use the technique will usually be taken by senior management. Thus an understanding of the potential and limitations of the technique is required if

correct implementation decisions are to be made. The Simulation Study Group (1991) found "there is a fear amongst UK managers of computerisation and this fear becomes even more pronounced when techniques that aid decision making are involved" This is combined with the fact that even those who do know how to use simulation become "experts" within a technically oriented environment. This means that those running the business do not fully understand the technique which could impact on their decision to use the results of the study.

Operational Involvement

Personnel involved in the day-to-day operation of the decision area need to be involved in the simulation project for a number of reasons. They usually have a close knowledge of the operation of the process and thus can provide information regarding process logic, decision points and activity durations. Their involvement in validating the model is crucial in that any deviations from operational activities seen from a managerial view to the actual situation can be indicated. The use of process maps and a computer-animated simulation display both provide a means of providing a visual method of communication of how the whole process works (as opposed to the part certain personnel are involved in) and facilitates a team approach to problem solving by providing a forum for discussion.

Simulation can be used to develop involvement from the operational personnel in a number of areas. It can present an ideal opportunity to change from a top-down management culture and move to greater involvement from operational personnel in change projects. Simulation can also be a strong facilitator of communicating ideas up and down an organisation. Engineers for example can use simulation to communicate reasons for taking certain decisions to operational personnel who might suggest improvements. The use of simulation as a tool for employee involvement in the improvement process can be a vital part of an overall change strategy. The process orientation of simulation provides a tool for analysis of processes from a cross-functional as opposed to a departmental perspective. This is important because powerful political forces may need to be overcome in ensuring departmental power does not prevent change from a process perspective.

The choice of simulation software should also take into consideration on-going use of the technique by personnel outside of the simulation technicians. For on-going use software tools need to provide less complex model building tools. This suggests the use of Visual Interactive Modelling tools which incorporate iconic model building and menu facilities making this type of simulation more accessible. There is also a need for training in statistical techniques for valid experimentation analysis. In summary the following needs are indicated:

- Knowledge transfer from technical personnel to managerial and operational staff of the potential application of simulation
- Training at managerial/operational levels in statistical techniques from companies and universities
- Training at managerial/operational levels in model building techniques from companies and universities

- Use of simulation as a communication tool between stakeholders in a change programme. The use of animation is useful.
- Use of suitable software, such as a visual interactive modelling system, to provide a platform for use by non-technical users.
- Incorporation of simulation in process change initiatives such a Business Process Reengineering (BPR).

Summary

This chapter has described the main activities in the design and implementation of a simulation model. These activities include the collection of data, the construction of a process map and statistical analysis of both input data and output statistics. An important element of a simulation project is a project report which documents the simulation model, presents the results of the analysis and suggests further studies. The organisational context of the use of simulation is discussed in terms of managerial and operational involvement in the simulation study process.

References

Black, K. (1992) *Business Statistics: An Introductory Course*, West Publishing Company, St. Paul.

Greasley, A. (2004), *Simulation Modelling for Business*, Ashgate Publishing Ltd., Hants.

Kelton, W.D., Sadowski, R.P. and Sturrock, D.T., (2007), *Simulation with Arena*, 4th Edition, McGraw-Hill, NY.

Law, A.M. and Kelton, W.D. (2000), *Simulation Modeling and Analysis*, Third Edition, McGraw-Hill, Singapore.

Muller, D.J. (1996), Simulation: "What to do with the Model afterward", *Proceedings of the 1996 Winter Simulation Conference*, eds. J.M. Charnes, D.J. Morrice, D.T. Brunner, and J.J. Swain, Society for Computer Simulation, San Diego.

Pegden, C.D.; Shannon, R.E. and Sadowski, R.P. (1995), *Introduction to Simulation Using SIMAN*, Second Edition, McGraw-Hill, Singapore.

Peppard, J. and Rowland, P. (1995), *The Essence of Business Process Re-engineering*, Prentice Hall, Hemel Hempstead.

Simulation Study Group (1991), *Simulation in U.K. Manufacturing Industry*, R. Horrocks (ed.), The Management Consulting Group, University of Warwick Science Park.

Welch, P.D. (1983), *The Statistical Analysis of Simulation Results*, The Computer Performance Modeling Handbook (ed) S.S. Lavenberg, Academic Press, New York.

5

Enabling Simulation – Simulation and Process Improvement Methodology

Introduction

This chapter investigates how a process-centred change methodology can be used as a framework for undertaking a simulation study and simulation can be seen as an important technique when undertaking process improvement.

A number of articles have been written concerning the tools and techniques of process improvement. Yu and Wright (1997) outline the need for tools to communicate process change initiatives at every level of the business. Kettinger et al. (1997) survey a number of techniques and suggest an approach for selecting techniques for a particular BPR project. Cheung and Bal (1998) state that a methodology for business improvement is really only as good as the tools and techniques that support it. An analysis is presented of tools categorised as paper based software-supported tools and software enabled tools. Majed and Zairi (2000) provide a review of BPR tools and techniques and reflect that BPR may be integrated into process management by pulling tools from a variety of process change approaches and building hybrid process design and implementation techniques.

When using simulation in the context of a process-centred methodology it is often referred to as Business Process Simulation (BPS). In general business process simulation is seen as a way of analysing dynamic systems that exhibit variability. Profozich (1998) outlines the disadvantages of using a static analysis tool such as a spreadsheet for analysing such systems, which requires the assumption that each process will operate on the average. There are however limitations to the simulation method. One disadvantage is the relatively large amount of time and cost needed to develop a model (Pidd, 1998). Further limitations can be apparent in the interpretation of simulation study results. For instance Fathee et al. (1998) state that as the complexity, randomness and variability within a business process increase the range of the prediction obtained from the simulation model becomes too wide for a decision base.

Business Process Simulation (BPS) (Tumay, 1996) has been traditionally used in a manufacturing context, but it is increasingly used in service organisations. Levine and Aurand (1994) describe the use of simulation to analyse an automated workflow system of an administration process and Verma et al. (2000) describe its use in redesigning check-processing operations. The use of simulation to assess the implementation of information systems (IS) in particular has been described by Giaglis (1999) and Giaglis et al. (1999) who outline the need to support IS evaluation by developing techniques for generating estimates of the organisational value of IS. Experimental methods such as simulation are suggested as capable of providing such estimates. Warren and Crosslin (1995) suggest simulation as a method of providing the justification necessary to support business decisions to redesign and make what are often massive investments in IT systems. A distinction should be drawn between studies of change to information system process design and change to the information infrastructure itself, in terms of elements such as the IS network configuration, telecommunications hardware and application software. Changes to these aspects are usually undertaken using network simulation software. Painter et al. (1996) argue that the process and infrastructure analysis can be integrated and present a methodology to achieve this.

Aguilar et al. (1999) indicate how BPS could provide support in a process centred management approach to change. Table 5.1. provides an adaption of that model. The model shows the use of simulation not only to predict the performance of the 'to-be' design before resources are committed, but also to use the technique to construct a model of the 'as-is' state in order to understand the process and measure the variation that takes place in key performance measures.

Table 5.1. Support provided by Business Process Simulation in process change

Phase	Step	Support provided by BPS
Assess 'As-Is'	Build and Communicate Process Map	Additional Support
	Measure and Analyse Process Performance	Major Support
Build 'To-Be'	Develop Future Process Design	Major Support
	Enable and Implement Future Process Design	Additional Support

Case Study 1 in this chapter provides evidence regarding the relationship between simulation and a process-based approach to change. The paper uses a case study regarding the implementation of an information system for road traffic accident reporting in a UK police force. Case Study 2 also relates Business Process Simulation (BPS), using the Aguilar et al. (1999) framework, to a process-centred management approach to change. Case Study 3 shows the use of a methodology for process improvement based on a study of process change with the Human Resources division of a police force.

Case Study 1
A Redesign of a Road Traffic Accident Reporting System Using Business Process Simulation

Introduction

A case study is presented of a proposed process change initiative to a police road traffic accident (RTA) reporting system. Two aspects of performance of the RTA system require particular attention. The need to speed process execution is seen as essential to provide a faster and more efficient service to vehicle drivers. In particular there is a need to provide UK government agencies, such as the Department for Transport, with accident statistics within a 4-week time period. The second aspect of performance which requires improvement is the need to reduce the relatively high staffing cost associated with the process. The total cost of traffic police staff is relatively high as their on-costs need to include the purchase and maintenance of a police patrol vehicle. There is also a need for extensive administrative support at locations across the area covered by the police force. This study outlines the construction of a business process simulation (BPS) to estimate potential cost savings of a computerised RTA system, through the use of allocation of staffing costs to activities.

The Road Traffic Accident Case Study

The case study investigates the computerisation of an RTA system at a police force level. The drivers behind the change are to both reduce the cost of the process and increase process execution speed. These objectives are to be met using digital technology to collect and transmit data to a central database, where in combination with information from a geographical information system (GIS) it can quickly provide internal and external agencies with relevant RTA information. Cost savings are to be made from the reduction in RTA officer hours in returning to base to collate and disseminate information and from the centralisation of administrative services leading to more efficient processes. The current and proposed process designs for actions leading from a road traffic accident will now be discussed.

The Current Road Traffic Accident Reporting System
The current process at the selected constabulary for the reporting and recording of accidents involves a police officer completing a number of paper-based forms following the report of an incident by the public. These forms are distributed to the traffic administration and data HQ departments for processing. The traffic administration section oversees the submission of witness evidence, either by post or in person and the collation of an abstract containing officer and witness statements for use by interested parties such as insurance companies and court proceedings. The data HQ section oversees mapping of the geographical location

of the accident that is used for road transport initiatives such as traffic calming and speed cameras. The main stages of the RTA process are now described.

Following the notification of a road traffic incident to the police by the public, a decision is made to attend the scene of the incident. It may be that for a minor incident the parties involved are instructed to pursue proceedings with their insurance companies and the police have no further involvement. If it is necessary to attend the RTA scene the officer travels to the location of the incident. After an assessment is made of the incident the officer returns to the station to complete and submit the appropriate paperwork. Three forms are used by a Police Officer attending a Road Traffic Accident (RTA). Form '54' is used for injury accidents and is triplicated on yellow, pink and white forms. The yellow and pink forms are forwarded by the officer to the witness proforma process and the white form is forwarded to the data HQ section for location mapping. A single form '55' is used for non-injury incidents which is filed unless further action is to be taken as a result of a dispute or claim, when it is then passed to the witness proforma process. In addition Pocket Book Entries (PBE) are taken when no official record is required but provide data that could be retrieved at a later date and transferred to the appropriate form. Amendment of forms may take place at a later date. Form '54' amendment forms contain yellow, pink and white sections for distribution. Form '54' (yellow and pink) and Form '55' amendments are scrutinised to see if changes require further action such as new witness statements. Form '54' (white) amendments are communicated to data HQ.

A location mapping process collates information and passes it to a local Council which provide a location grid reference from sketches and location information provided by the officer who attended the accident scene. The data is collated, sorted and then forms are mapped in batches by entering location codes on new forms. If all the necessary information is not available a memo is sent to the officer for further information and the process is repeated. The information is then sent electronically to the local council who return the required geographical location details.

The witness proforma process obtains accident witness and driver information and places it on a proforma sheet. If a witness is identified their details are taken and a proforma is sent to them. If a fatal accident has occurred then the officer obtains further details in person at a later date. If the proforma has not been returned after three weeks a reminder letter is sent to the witness. If there is still no response from the witness and the information requested is required for further proceedings, then an officer will obtain the statement in person.

The abstract preparation process collates and checks documents associated with the RTA process to ensure all the data needed has been received. A decision is made at this stage if further action is required after reviewing the evidence collected. If further action is required a number of forms are collated. If at this stage no prosecution is to take place, a letter informing the driver of this decision is sent. If a prosecution is to take place the officer will write an abstract, summarising the details of the case. If a court case is scheduled and a not guilty plea has been entered then the officer will be required to attend the court proceedings in person. Otherwise this is the end of the involvement of the officer.

The Proposed Road Traffic Accident Reporting System
In the proposed computerised RTA reporting system the attending officer completes paper-based forms as before but this information is promptly converted to digital form using a document image processing (DIP) system. This is achieved by a combination of image capture and data recognition through a facsimile link. Data recognition systems, such as optical character recognition (OCR) are used to process information that is entered in a structured format, such as options selected using a ticked box format. Image capture is used in the following ways. Documents are stored as images to enable input bureau staff to validate the OCR scanned data. Images which cannot be interpreted by data recognition software, such as hand drawn sketches of the RTA scene are stored for later retrieval. Images of text such as officer written notes can be entered by input bureau staff, saving officer time. Once in digital format the documents can be delivered electronically preventing data duplication and enabling faster distribution. Physical documents are held in a central repository for reference if needed.

Location details are currently based on a written description of the RTA by an officer which leads to inconsistent results. The current location description by the officer is usually acceptable for city incidents where nearby street intersections and other features can be used to pinpoint a location. However on long stretches of road it is often difficult to pinpoint an exact spot. This is important because of the need to accurately pinpoint areas with high accident rates for road safety measures (e.g. road humps) and speed camera placement. Further inaccuracies can also occur when the officer description is converted by the local council using an Ordnance Survey (OS) map grid reference which is only accurate to 200 yards. In this proposal each officer is issued with a portable digital map on which to indicate the RTA location. This information is transmitted by a mobile link to a geographical information system (GIS) (Wiley and Keyser, 1998) which provides accurate location analysis of both injury and non-injury incidents using the geocode system (Radcliffe, 2000). The geocode system is a network of grids covering the UK which allow a location to be assigned within a $10m^2$ area. The GIS system will combine the accident location analysis with data relating to the location of pelican crossings, traffic lights, street parking and anything else that might contribute to accidents or affect schemes being proposed. Along with data on details on road conditions at the time of the accident this information will help determine a prioritised list of road safety improvement measures.

The Road Traffic Accident Business Process Simulation

The implementation of the business process simulation will be described in four steps based on the model of using business process simulation to support process centred change presented by Aguilar et al. (1999).

Step 1: Build and Communicate Process Map
Data requirements for the simulation model can be grouped into two areas. The first area of data is required in order to construct a process map which describes the logic of the model (i.e. how the process elements are connected) and decision

points within the process. Decision points can be modelled by a conditional rule based method or by a probability distribution. A process map of the proposed RTA reporting system is shown in Figure 5.1. Probability distributions for decision points, such as the proportion of injury and non injury events are derived from the sample data and take the form of a percentage. The second area of data required for the simulation model is for additional elements such as process durations, resource availability schedules and the timing of RTA occurrences. In this case probability distributions for process durations are derived from the sample data. In general a triangular distribution has been used for process durations which require minimum, mean and maximum parameter values. Resource availability, in terms of a police officer attending the RTA, is assumed to be infinite as an RTA incident is treated as an 'emergency situation' and if the designated officer is unavailable an alternative officer is found. Over a period of 6 years there had not been an incident when no officer could be found, when required, to attend an RTA scene. In terms of the timing of RTA occurrences a three-month sample of road traffic accident (RTA) incidences was collected for the study. Although seasonal variations will occur in RTA patterns, the model is used to compare before and after redesign scenarios and not for forecasting purposes.

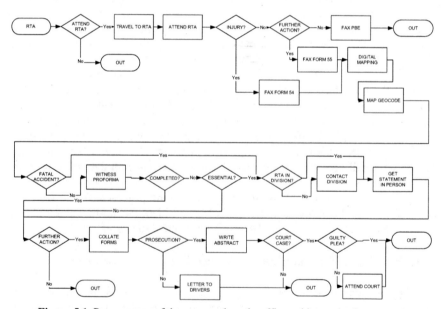

Figure 5.1. Process map of the proposed road traffic accident reporting system

Step 2: Measure and Analyse Process Performance
The model was built using the ARENA™ discrete-event simulation system (Kelton et al., 2001) that incorporates a template of shapes called blocks that are placed on the computer screen and connected to represent the logic of the model. The user can then mouse click on each block and a pop-up menu allows the entry

of parameters such as process durations. The main elements that make up a model are the CREATE, PROCESS and DECIDE blocks. The CREATE block generates the arrival of people (service applications), physical components (manufacturing applications) or information (information system applications) into the system. These are generically called entities in simulation terminology. It is necessary to define the rate of arrival of entities in the system by defining the time between arrivals (interarrival rate). This was achieved using sample data on the timing of the occurrence of RTA incidences. The PROCESS block is used to delay an entity for a predetermined time period (to represent the time taken to attend the RTA for example). It is also used to allocate resource time (e.g. traffic officer time) to a process. The DECIDE block is used to define decision points within the process. The entity can leave from 2 or more outputs from the block. For a 2-output decision the routes are labelled true and false. These decision options can be implemented as either a percentage chance for each decision route (e.g. 70% entities follow the true route, 30% false) or by a rule-based decision (e.g. IF x THEN true route, else FALSE route).

Before experimental analysis of the model can begin it is necessary to ensure that the simulation provides a valid representation of the system. This process consists of verification and validation of the simulation model. Verification is analogous to the practice of 'debugging' a computer program. In this case the first check was to run the simulation model animation at a slow speed and observe the progress of the entities through the system to check for any logic errors. Then a number of test runs were undertaken and the results of the simulation noted and checked against real system performance and common-sense deductions. A verified model is a model which operates as intended by the modeller. However this does not necessarily mean that it is a satisfactory representation of the real system for the purposes of the study. This is the purpose of validation. In this case the results from the current RTA reporting system model could be compared against historical data of the actual system. A decision is made if model behaviour is close enough to the real system to meet the objectives of the study. Unlike verification, validation is a matter of judgement which involves a trade-off between the accuracy of measurement required and the amount of modelling effort required to achieve this.

Step 3: Develop Future Process Design
Because of the probability distributions used for RTA events, process times and decision points, the output measures of the simulation vary each time the simulation is run. Therefore it is necessary to run the simulation multiple times and form a confidence interval within which the average of the measure should lie. In this case the simulation was run 10 times for a simulated 28 days for the current and computerised systems. The amount of road traffic officer time in hours was noted for each run for each scenario and a confidence interval calculated at a 95% level. In this case the confidence intervals do not overlap and so there is a significant difference in the results (Robinson, 1994a). In other words the difference in the result for the current and computerised systems is not due to random variation alone but due to an actual difference in performance. The results show that the mean officer hours required to undertake all the tasks associated

with the RTA process is 2049 hours under the current system and 1750 hours under the computerised system. With an on-cost pay rate for road traffic officers estimated at £17.83 per hour this converts to a cost saving of £5331 per 28 day period or approximately £63,971 per annum.

Step 4: Enable and Implement Future Process Design
The savings outlined in Step 3 are dependent on a number of changes in process design. The need for officer time for transcribing and updating notes will be minimised by the use of Optical Character Recognition (OCR) transcription of officer notes by data centre personnel. A centralised data store will also save officer time by quicker storage and retrieval of information for the proforma and abstract preparation process. In addition the location mapping exercise will be simplified by the use of portable digital maps from which officers can indicate the RTA location. Process time will be speeded by workflow automation software which will prompt for timely response to requests for information in the witness proforma and abstract preparation processes. The use of a single point of contact for all data submissions and information requests will also reduce process execution time by eliminating search and delivery delays associated with paper records. The IS system will also have the benefit of improved data accuracy with a single database of all information and location analysis through the use of geocodes.

The simulation study focussed on savings made on the front-line road traffic officer staff but substantial savings can also be made by the centralisation of the traffic administration units. These units are currently located at a divisional level, which is a geographical subdivision of the police force area. This is necessary so that paperwork can be processed from officers returning to their local stations. However with the use of digital transmission of RTA information the geographical location of the administrative support can be centralised at a Force level. This can lead to less staff needs due to a centralised automation of processes through workflow and database technologies. The demand on separate divisions would also be aggregated at a corporate level leading to more efficient staff utilisation.

Discussion

The case study has presented the technique of business process simulation used to facilitate business process change enabled by the introduction of an information system. What the simulation was able to do was to both demonstrate how the new process would execute and quantify savings in officer time. Demonstration of the operation of the new design using the animation display reduced uncertainty in how the new system would operate. Observation of performance measures provided by the model helped to secure an acceptance of the need for change by demonstrating the increased performance of the proposed system. In particular information provided by the model of staff cost savings quantified the benefits of change in terms of traffic officer on-costs.

A major issue in the organisation at that time was a lack of confidence in the introduction of IS systems due to the fact that previous IS projects had often been delayed or over-budget. Gulledge and Sommer (2002) outline the importance of aligning information systems with business processes in public sector organisations and although the advantages of the computerisation of the RTA reporting system are evident in increased efficiency, in reality the change requires a number of complex issues to be addressed. These include the integration of the new technology systems with legacy systems (for example the integration of database systems), the management of a reduction in staffing levels and the movement of staff from divisional to a centralised traffic administration unit. While not contributing directly to the implementation of these organisational changes the simulation was able to provide a level of confidence both in the operation of the new process design and quantification of potential benefits in terms of improved efficiency. Both of these factors provided an impetus to initiate the proposed changes.

Further studies could extend this analysis with a study of the information infrastructure required. Thus information on both process and infrastructure costs could be combined to form the basis of an investment appraisal tool for the introduction of Information Systems. A model of the information infrastructure would also have the advantage of assessing the feasibility of the project in terms of determining a specification for hardware such as communication devices and database systems.

Case Study 2
Using Business Process Simulation Within a Business Process Reengineering Approach

Introduction

A case study is presented of changes made to the operation of a custody process in a UK police force.

The Custody of Prisoner Process Case Study

A thorough review of the Information Systems used within a police force was conducted and many shortcomings were found with the processes involved. Problems included too many Police Officers performing jobs that could be automated or performed by civilians and computerisation of processes at a local level leading to inefficient information handling. It was envisaged that efficiency gains could be achieved by undertaking detailed process reviews on a force-wide basis. To undertake this review a business-led process review team was set up. Team members were chosen with a mix of operational, administration and information systems experience. The team was not part of the I.T. function but reported directly to senior management. The previous information systems review had identified a number of priority areas for investigation including custody, case preparation, accounting and personnel. It was decided to focus on the custody operation first because it is legally bound by nationally structured procedures. This would allow the team to gain experience in redesign before moving on to more loosely defined processes. The custody process under investigation includes the arrest process, from actual apprehension of a suspect, to processing through a custody suite (which contains booking-in, interview and detention facilities) to a possible court appearance. A number of different police roles are involved in the custody process including arrests by a Police Constable (PC), taking of personal details by a Custody Officer and supervision of persons in detention by a Jailer. The first objective of the study was to identify staff costs involved in arresting a person for a particular offence under the current design. The second objective of the study was to predict the change in utilisation of staff as a result of re-designing the allocation of staff to activities within the custody process. The case will be described using the steps outlined in Table 5.1.

Step 1: Build and Communicate Process Map
A process map (Figure 5.2) was constructed after discussions with police staff involved in the custody process. As this process was legally bound, documentation on the order of processes and certain requirements such as meals, visits and booking-in details was collated. The main activities in the arrest process are shown in the process map. Each decision point (diamond shape) will have a probability for a yes/no option. The first decision point in the arrest process is whether to conduct a search of the location of the arrest. For all decisions during the arrest process an independent probability distribution is used for each type of arrest (e.g. theft, violence, drugs) at each decision point. The staffing rank required for each process is indicated above the process box. Personnel involved in the arrest process include the PC, Custody Officer, Jailer and Inspector. The role of each rank is indicated on the process map for each activity.

The process map would normally provide the basis for a simple analysis of the process or form the basis of a workflow model. However in this case it provides information for the logic of the simulation model (i.e. how the process elements are connected). In the terminology of a simulation study the process map constitutes the conceptual modelling stage. In addition to the process logic contained in the process map, the simulation requires additional data in order to undertake a dynamic analysis of the system, rather than the static model represented by the standard flowchart type process map.

A sample of process durations should be collected from which a statistical distribution of the duration can be derived. Statistical techniques are used to ensure a suitable distribution and parameters are chosen. In this case a process duration distribution was derived for each type of arrest. To model the variability in process times a number of estimated times were collected from a number of custody officers. In addition videotapes of the booking-in and interview procedures were viewed and timings taken.

The demand level of each arrival type is estimated from which a statistical distribution for the 'time between arrivals' is usually determined. In this case data was gathered on the timing of arrests over a period of time from information contained within booking-in sheets. From this data the demand pattern was analysed for a typical police station during a day and for each day in a week. As

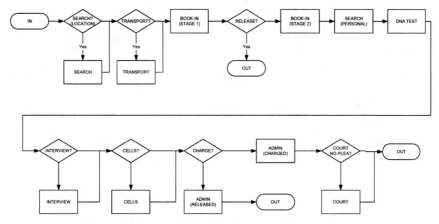

Figure 5.2. Process Map of the Custody of Prisoner Process

expected demand fluctuated both during the day and differed between days in the week. It was decided to use historical data to drive the model due to the nature of the demand, with a number of arrests occasionally occurring simultaneously (as a result of a late-night brawl for example!), which would be difficult to reflect using an arrival distribution. Additionally the use of actual data would assist in model validation as model performance could be compared to actual performance over the period simulated. This approach was feasible because the focus of the study was on the investigation of the comparative performance between different configurations of custody operation rather than for use as a forecasting tool.

Most processes will consume resource time, which may be of a machine or person, and these resources need to be identified and their availability defined. If the resource is not available at any time, then the process will not commence until the resource becomes available. In this case the main resources are Police Constable, Custody Officer and Jailer.

Step 2: Measure and Analyse Process Performance
In order to measure and analyse process performance a business process simulation was constructed using the ARENA™ simulation system (Kelton, 1998). This system uses icons (representing processes) that are placed on a screen area. The icons are connected by links to represent the logic of the process. Process duration and resource allocation to processes is made by double-clicking a process icon and entering data in the pop-up menu dialog. An animated display is constructed by using an in-built graphics package, which permits the construction of background (static) elements such as the process layout schematic and animated (dynamic) elements which move across the screen. Figure 5.3 shows the custody display that consists of representations of the main custody area facilities (i.e. booking-in desk, interview rooms and cells) and the dynamic elements (i.e. arrested persons and police staff) that move between the custody facilities.

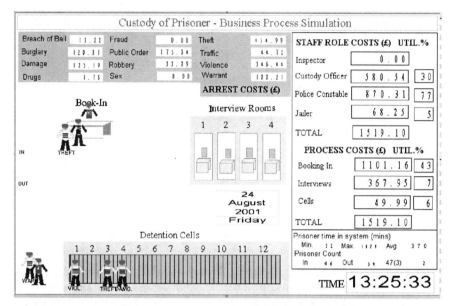

Figure 5.3. Custody of Prisoner – Business Process Simulation Display

Before the model results are recorded model behaviour must be checked to ensure that the model is providing valid results. Verification is analogous to the practice of 'debugging' a computer program. This is accomplished by techniques such as a structured walkthrough of the model code, test runs and checking of the animation display. Validation is about ensuring that model behaviour is close enough to the real-world system for the purposes of the simulation study. To achieve this the model builder should discuss and obtain information from people familiar with the real-world system including operating personnel, industrial engineers, management, vendors and documentation. Also the technique of sensitivity analysis, to test the behaviour of the model under various scenarios and compare results with real-world behaviour, can be used.

Once the model had been validated, it was run over a set time period and results collected. At this stage the model is simply reproducing the behaviour of the current process. This 'As-Is' model provided a visual representation of the whole process, which was important to provide a consensus that the model provides a convincing representation of the process. Demonstration of the model between interested parties provided a forum for communication of model behaviour and helped identify any anomalies. In this study the aim of the model was to identify the main sources of cost in the system and thus provide strategies which would enable cost to be reduced. At present a budget–based approach meant that costs were not allocated to activities within the process. Each time a process was activated in the model by an arrested person being processed through the custody system a cost was calculated by multiplying the activity duration by the cost per time unit for the resource allocated to that process. The initial analysis aimed to identify the cost incurred for each type of arrest incorporated in the model. The results of the analysis are shown in Table 5.2.

Table 5.2. Arrest cost by arrest type

Arrest Type	Average Number of Arrests per Month	Average Cost (£)	Average Cost per Arrest (£)	Percentage of Costs	Percentage of arrests
THEFT	266	18600	70	33%	28%
BURGLARY	85	7172	84	13%	9%
VIOLENCE	104	6755	65	12%	11%
DAMAGE	91	5522	61	10%	9%
WARRANT	189	5349	28	9%	20%
DRUGS	43	3522	82	6%	4%
PUBLIC ORDER	63	3083	49	5%	7%
BREACH OF BAIL	49	2695	55	5%	5%
SEX	15	1362	91	2%	2%
TRAFFIC	20	983	49	2%	2%
FRAUD	19	972	51	2%	2%
ROBBERY	12	822	67	1%	1%
	958	56837			

The cost for each arrest type is a function not only of the number of arrests but the likelihood of an arrest leading to interview, detention and court procedures. In this case relatively trivial theft offences (usually involving children shoplifting) are causing a heavy workload and thus a high cost. Thus a driving factor behind the overall cost structure has been identified. A possible way of reducing cost could be to decrease the theft activity through crime prevention activities for example. Greasley (2001) provides more details of the use of the simulation in conjunction with the Activity Based Costing (ABC) technique.

Step 3: Develop Future Process Design
In this case the 'to-be' model was used was to explore the reduction of staffing cost by re-allocating staff roles to processes. By estimating resource costs, in this case staff wages, it was possible to estimate the effects and feasibility of proposals to re-allocate and civilianise staffing duties within the custody process. Greasley and Barlow (1998) outlines the effect of civilianisation of certain staffing activities.

Step 4: Enable and Implement Future Process Design
Once a future process design has been decided, the simulation helps to implement this change in a number of ways. The graphical display provides an excellent tool with which to communicate the process to stakeholders such as management, customers and the workforce. The 'before' and 'after' graphic displays can be used to show how the changes will affect the process in practice. Also the display of the new process design can be utilised to train staff in the new operation and provide then with an overview of process behaviour which enables them to see the relationship between a particular process activity and the overall process behaviour.

The graphics are complemented by performance measures to quantify before and after performance and thus demonstrate potential improvement. In the custody of prisoner case, measures of staff utilisation were important in demonstrating the feasibility of the re-allocation of tasks between staff. The figures also quantify potential savings in the utilisation of police staff time, enabling plans to be made for the re-allocation of staff time to other duties. In the analysis of workload in terms of arrest types, cost was used as a measure of the aggregated staff resource allocated to each arrest type that is serviced by the police. The simulation analysis could take account, not only of the number of arrests of each type, but the variable number of processes (e.g. interview, court appearances by PCs) that each arrest triggered. The modelling of the complexity of interdependencies between arrest processes and the variability of arrest demand allowed an accurate picture of where cost/effort was being spent. These figures provided an impetus to focus expenditure on programmes, such as crime prevention, which could have a substantial effect on cost/effort and could free resources to provide an improved overall performance.

Discussion

Authors have differing opinions of the usefulness of business process simulation or simulation software in the context of business process reengineering projects. For example Petrozzo (1994) outlines how BPS helps understanding of process dynamics whilst Peppard (1995) is wary of the use of simulation analysis in the redesign stage due to the potential time and cost involved in building the model. This section discusses the benefits and limitations of BPS in the context of the case study project and also outlines what contribution a BPR approach can make to the successful use of the simulation technique.

Benefits of BPS for BPR
The main advantage of the BPS analysis is that it allowed the incorporation of variability and interdependence factors in order to obtain an accurate outline of process performance. For instance the cost per arrest type could not be determined from the process map alone because this gives no indication of the likelihood that a particular sequence of events (e.g. interview, cells, and court appearance) would occur for a particular arrest type. The simulation could predict process performance along a number of measures such as lead-time, resource utilisation and cost. In this case a cost element was used to estimate expenditure by staff role, process and arrest type. Efforts were then made to reduce cost across these three aspects by civilianisation, process redesign and crime prevention activities respectively. Once built, the BPS allowed analysis of many potential new designs through its 'what-if' capability with little extra effort. For example a number of booking-in process redesigns could be analysed in advance using the model, before implementation.

Another benefit was provided by the visual animated display which provides a communication forum to both validate the model and to explain the operation of redesigned activities and their role in overall process performance to staff. Bell et al. (1999) report on the generally high level of support and interest in visual interactive models by decision-makers. In some instances it may be that due to a lack of input data the BPS is used, not for a detailed quantitative analysis, but to facilitate discussion and ideas by the use of the visual interactive display (Robinson, 2001).

Limitations of BPS for BPR
A major barrier to many organisations in using BPS is the preparation needed in the successful introduction of the technique to the organisation. In order to overcome this Harrington and Tumay (2000) provide a 4-phase model which provides a detailed plan for a successful introduction and continued use of simulation in the organisation. A particular emphasis is placed on the need for a simulation sponsor who disseminates information of the technique, as well as the training of users on simulation software.

The potential that process mapping will lead to too much emphasis on operationalising existing processes, rather than conceptualising a new design has been recognised (Cotoia and Johnson, 2001) and BPS could be said to increase the scope for over-analysis. However, it should be recognised that when estimating the amount of resource required to construct a BPS, that there should be no attempt to model every aspect of the area of study, but the level of detail and scope of the model should be judged according to the study objectives (Robinson, 1994b). Thus building a sophisticated model must not become the objective of the exercise, the model should be built with just enough detail to provide information on which to make decisions.

One limitation of BPS in the context of BPR projects is that the BPR team must be careful not to create a 'to-be' design based solely on 'tweaking' the 'as-is' simulation model. 'Simulation is most useful in comparing 'as-is' and 'to-be' models, and validating and ensuring the completeness of the 'to-be' process model. Beyond this simulation has limited ability in creating a 'to-be' model.' (Levas et al. 1995). Thus simulation will not create a new design and design ideas should not be constrained by the complexity of changing the model to simulate the new design. Design ideas should drive simulation design, not the other way around.

Finally Fathee et al. (1998) note that simulation is most useful for the analysis of stable business processes and less useful for dynamic systems that do not reach equilibrium. This may point to potential difficulties in statistical analysis of service systems that are less likely to reach equilibrium than a manufacturing process.

Discussion of BPR for BPS
While BPS can bring the above facilities to a BPR project, BPR in turn can provide a framework for BPS implementation. Business Process Simulation and process based change methodologies such as BPR have a number of characteristics in common. For instance they both use as a basis for analysis the process map that shows the interrelationships between activities within a process. This approach facilitates communication of the whole process, across organi-

sational boundaries using the visual medium of the process chart. Thus both BPR and BPS have a process orientation and use the process map as an analysis tool.

Robinson (2001) states that BPS is not a methodology in itself but a technique that can be used to support a chosen methodology. Thus the BPR methodology could provide direction for the design of the 'to-be' process and ensure that the BPS study objectives are linked to the strategic objectives of the organisation. The importance of the link between strategy and business process redesign has been recognised by Harrison (1998) 'Focus on processes that do not affect the firm's strategic future misdirects scarce resources into doing the wrong things right, or into reengineering processes in a way that is insensitive to their competitive contribution.'

Another aspect that a process-based change approach can bring to a BPS project is the need to understand human factors such as culture, motivation and leadership style in the implementation of change (suggested by authors such as Buchanan (1998) and Campbell and Kleiner, 1997). For example Buchanan (1998) rejects a "blank sheet" approach to reengineering as a result of analysis of process change in a NHS hospital. Here a "fresh start" approach to reengineering is seen as impractical in an organisation with other organisational priorities (e.g. the implementation of change triggered by shifts in government policy) and barriers to change rooted in the sector's history and culture. However a process-based change approach which takes these organisational context issues into consideration is seen as potentially consistent with a creative and participative approach to organisational development. In this case a process-based methodology can provide a framework to ensure realistic design scenarios are analysed by the simulation and assumptions made in the model building process (for example regarding the allocation of work roles) take account of the overall organisational context within which the change is taking place.

Future Research

Although simulation software is increasingly sophisticated and user friendly, a manufacturing/engineering background has limited its use in the service sector, where it would seem to be an ideal tool for analysing customer processing applications. The use of process-based change methods provides an opportunity to widen the use of simulation and utilise its ability to analyse variability factors leading to a more accurate view of process performance and allow different designs to be tested reducing the risk of change. The custody study is intended to show how BPS can be incorporated into a process change methodology. However in interpreting the results from the study, the unique culture of the police force and characteristics of the custody process need to be considered. The custody process is based on a number of rules, often laid down in legislation, covering the arrest process. These facilitate the model building process, but other service processes may be more ill-defined in such areas as the allocation of staff to work roles and the process flow of people through the system. Thus more case studies showing the use of simulation in a variety of organisational contexts in the service sector are required. These need to show the benefits, not just of the statistical analysis, but the relationship between BPS and a process-based approach to change.

Case Study 3
Process Improvement Within a HR Division at a UK Police Force

Introduction

This paper presents a case study of an organisation undertaking a process improvement exercise using a variety of process techniques and tools. The case study involves a reengineering project which has taken place within a Human Resources (HR) division of a UK police force. It was recognised that it was important to link the change programmes initiated by the reengineering effort to strategic priorities derived from a range of stakeholder interests, rather than being budget driven through a range of financial indicators. In public sector organisations such as the Police there will be a need to reconcile the multiple objectives of stakeholders such as the government, employees and victims of crime. A balanced scorecard approach was undertaken in order to achieve this. The next step was to conduct a process mapping exercise in order to identify the activities and the relationship between activities in the HR division. The next stage required the identification of process elements which should receive priority for the improvement effort. This article introduces the use of a scoring system which prioritises processes for improvement according to their 'effect on performance' and 'amount of innovation required'. The scoring system provides management with a starting point for aligning the improvement process with the strategic objectives of the Force identified in the balanced scorecard analysis. The next stage of the study is then to redesign the processes to improve performance. This was undertaken using the rules of Eliminate, Simplify, Integrate and Automate (ESIA). A performance measurement system is then shown which links the operational performance of business processes to strategic targets over time.

The HR Division Case Study

The nature of the police force as a service industry means that a central task is the management and deployment of human resources. As part of a reengineering study covering all aspects of Police duties an investigation of the HR division and its I.T. infrastructure was undertaken. One particular area selected for redesign was the sickness and absence process located in the HR division. This is of particular interest as the issue of illicit absenteeism from work in the UK Police Service has recently attracted publicity. It has been estimated that it was costing police forces in England and Wales £250million a year, losing an average of 11 working days per officer annually (Sheehan, 2000).

The following steps were undertaken in the process improvement exercise.

1. Derive the critical success factors
2. Process mapping

3. Identify processes for improvement
4. Process redesign
5. Measure performance

Johansson et al. (1993) presents a BPR methodology of Discover, Redesign and Realise. Relating these steps to that methodology, Step 1 would fall within Discover, Steps 2–4 within Redesign and Step 4 within Realise. The steps are now outlined in more detail within the context of the process improvement exercise in the HR division.

Step 1 – Derive the Critical Success Factors
The first step in deriving strategic objectives of the improvement programme was to conduct a stakeholder analysis. The stakeholder approach states that an organisation needs to pursue performance across a number of perspectives and take a long-term view in order to satisfy the needs of all interested parties (i.e. stakeholders) in the business (Doyle, 1998). For example the single-minded pursuit of shareholder value can erode the trust of employees and customers as shareholder interests override other factors. The main task of management is to reconcile the diverging and partly conflicting interests of stakeholders such as employers, customers, managers, shareholders, creditors and government and community agencies.

The function of the Police can be summarised as to uphold law and order, enforce the law, prevent crime and apprehend criminals (Farnham and Horton, 1996). Under the stakeholder approach the interpretation of the relative importance and resource given to the above is due to the trade-off in power and interest between a number of stakeholder groups. The main stakeholder groups specific to the Police include the Home Secretary who sets down objectives and performance targets at a national level. He or she is responsible for efficiency and effectiveness, but not for establishment or terms and conditions of service. Local Police Authorities (LPA) approve a costed plan to national and local objectives, monitor and publish performance and appoint the Chief Constable. LPA's can be seen as the 'purchasers' of the Police Service for their area. The Chief Constable of the force is responsible for plans, budget, staffing, buildings, suppliers, equipment and is the de-facto employer of all police staff. The Chief Constable can be seen as the 'supplier' of the Police Service for their area. Other stakeholder interests include the employees, which may be employed directly or indirectly by the Police and who may be uniformed or civilian. To operate effectively the Police also require the consent and support of the public.

The wide range of stakeholder groups means that the strategic control system needs a balanced set of performance indicators to reflect the views of a diversity of national bodies which set standards and the local community. Critical Success Factors (CSF) can provide a guide to determine 'what needs doing well' in order to implement a strategy and fulfil the organisational vision. The critical success factors can be placed within the four perspectives of the balanced scorecard (Kaplan and Norton, 1996) which provides a balanced set of performance indicators to reflect the views of the wide range of stakeholder groups involved. A balanced scorecard can be constructed at the organisational or departmental level at which a focused strategy can be adopted. 'A department should have a balanced scorecard if that organisational unit has (or should have) a mission, a strategy, customers (internal or external) and internal processes that enable it to accomplish its mission and strategy' (Kaplan and Norton, 1996). The critical success factors for the HR organisation were based on the strategic plan developed by the police force at a divisional level. The objectives were then passed to the heads of the various departments within the HR division (e.g. head of personnel, head of training) for discussion. The CSFs are shown in Table 5.3.

Table 5.3. Critical Success Factors for Human Resource Division

Perspective	Critical Success Factor
Innovation and Learning	*Increase Individual Performance* This CSF needs positive improvement and measures to improve individual performance. E.g. Health and welfare of staff, rewards/penalties, attract quality applicants, retain staff, staff relations, development and training of staff.
Business Process	*Increase Effectiveness of Strategic Management* This CSF will provide accurate management information on which to base strategic development and policies and decisions. E.g. improve service to public, develop policies, support and advise on organisational change, empowerment.
	Improve Staff Communications This CSF will ensure that staff are better informed of and are aware of new legislation, policies and procedures, where to obtain relevant management information and expert advice and set up procedures to enforce policies particularly those mandated by legislation or otherwise.
Stakeholder/ Customer	*Meet Legislative Requirements* Failure to meet this CSF will result in penalties including financial and dismissal e.g. EC Regulations, Health and Safety, Working Time Directive, Police Regulations, Employment Legislation etc.
	Increase Effectiveness of Force in its Delivery to External Customers This CSF will require HR to provide accurate management information to the force on which the force can base both strategic and tactical responses to the needs of external customers and agencies.
	Increase Effectiveness in Delivery to Internal Customers This CSF will require HR to provide accurate management information and professional advice to heads of departments and line managers (all levels) on which heads of departments and line managers can base both strategic and tactical responses to the needs of their own staff and other internal customers.
Financial	*Improve Value for Money (of HR to the Organisation)* This CSF will require HR management to develop strategic and tactical plans and policies to guide force strategic development in HR aspects and make most effective use of resources and systems to provide a high quality service

Step 2 – Process Mapping
In order to identify activities within the sickness and absence process the technique of process mapping is used. Process Mapping involves a study of how activities link together to form a process. The technique involves interviewing personnel and observation of the relevant process which provides information which is used to draw a process map. The analysis shows the interrelationships between activities and identifies the elements and roles involved in process execution. Harrington (1991) outlines a number of mapping and flowcharting techniques. In this case the process maps were implemented using Microsoft VISIO software.

Step 3 – Identify Processes for Improvement
Once the process mapping has been completed it is necessary to prioritise the process elements which will be allocated resources for improvement. The identification of the relevant business processes for improvement can be undertaken using a scoring system such as the performance/importance matrix (Martilla and James 1977) on which processes can be plotted in terms of how well the organisation performs them and how important they are. Slack and Lewis (2002) outline a model in which prioritisation is governed by importance to customers and performance against competitors. Crowe and Rolfes (1998) outline a decision support system for identifying which processes to select for improvement taking into consideration the organisation's strategic objectives.

The following example presents a scoring system developed in conjunction with the Force. The system consists of a two-dimensional marking guide based on the impact of the process on the critical success factors determined in the balanced scorecard review and an assessment of the scope for innovation (i.e. the amount of improvement possible) to the current process design. Processes which are strategically important and offer the largest scope for improvement are prioritised under this model. The marking guide marks each process on a scale of 0 to 5 against 2 measures, IMPACT – The extent to which the achievement of the CSF depends on the process and INNOVATION – The extent of the change required to the process in order to meet the CSF. The marking guides for each measure are shown in Table 5.4 and Table 5.5.

Table 5.4. IMPACT (External Perspective) Marking Guide

Mark	IMPACT (External Perspective) Marking Guide
0	This individual process has minimal or no effect on the individual CSF
1	This individual process is dependant on another process, in order for it have an effect on this CSF
2	This individual process has a marked influence on this CSF
3	The individual process has substantial impact on whether another process can maximise its beneficial effects on this CSF
4	The individual process has substantial influence on this CSF
5	The individual process is a critical part of being able to achieve the individual CSF

Table 5.5. INNOVATION (Internal Perspective) Marking Guide

Mark	INNOVATION (Internal Perspective) Marking Guide
0	This process cannot be improved for this CSF
1	This process achieves its objective but could be improved even further.
2	This process achieves its objective but could be improved by review of both automation and process improvement.
3	This process does not effectively achieve all its objectives and could be improved by review of both automation and process improvement.
4	The process exists and functions but needs substantial alteration to meet its objectives.
5	The process either does not exist or only partially exists and fails to meet any objectives.

In terms of the balanced scorecard the IMPACT measure relates to the achievement of the CSF from the stakeholder and financial (external) perspectives of the balanced scorecard. The INNOVATION measure relates to the amount of change required from the learning and business process (internal) perspectives (Figure 5.4).

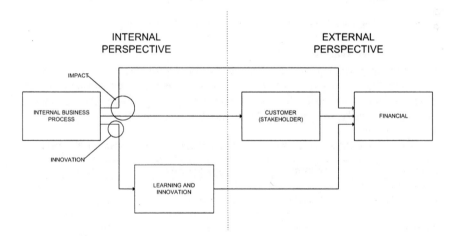

Figure 5.4. Relationship between the scoring systems and the balanced scorecard perspectives

Each process element is scored (0–5) against each CSF for the impact and innovation measures. The score for each measure is multiplied to provide a composite score (0–25) for each CSF. An overall composite score for each process is calculated by adding the composite score for each CSF. A spreadsheet sort by composite score identifies a priority list of processes for improvement. Figure 5.5 shows a spreadsheet implementation of the scoring system for the sickness and absence process by process order.

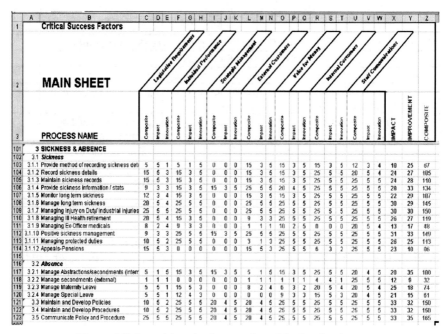

Figure 5.5. Spreadsheet implementation of the Scoring System

Step 4 – Process Redesign
Once suitable processes have been identified for improvement the next stage was to redesign these processes to improve their performance. Peppard and Rowland (1995) provide a number of areas of potential redesign under the headings of Eliminate, Simplify, Integrate and Automate (ESIA) (Table 5.6).

Table 5.6. ESIA areas for potential redesign (Peppard and Rowland, 1995: 181)

Eliminate	Simplify	Integrate	Automate
Over-production	Forms	Jobs	Dirty
Waiting Time	Procedures	Teams	Difficult
Transport	Communication	Customers	Dangerous
Processing	Technology	Suppliers	Boring
Inventory	Problem areas		Data capture
Defects/failures	Flows		Data transfer
Duplication	Processes		Data analysis
Reformatting			
Inspection			
Reconciling			

In the case of the 'sickness and absence' process this involved integration of the present 30 separate process maps into a single process. A number of duplication of activities was found primarily due to functions included with the sickness process being based both at Headquarters and Divisional level. The redesign improved performance both internally and externally. Internally process efficiency was increased with an estimated saving per sickness/absence event of between 45 minutes to 60 minutes of staff time, representing a significant resource saving which can be re-deployed to improve performance. External process effectiveness has been increased by increasing the speed of follow-up checks to absent officers and new innovations such as requiring earlier checks by the Police doctor.

Step 5 – Measure Performance
Neely et al. (1995) provide a survey of performance measurement systems and provide as a research agenda the need for organisations to integrate their performance measures into their strategic control systems. McAdam and Bailie (2002) confirm through a longitudinal case study approach that performance measures linked to strategy are more effective. In this case in order that progress is maintained towards strategic objectives it was considered important to relate performance of the process at an operational level to strategic targets. The 'Sickness and Absence' process is related to the CSF identified in the balanced scorecard initiative of increasing individual performance. At a strategic level the measure of staff productivity was chosen with a target to increase availability of police officers by 5%. In order to meet this strategic target, measures and targets are needed at an operational level. These are derived both from the strategic measure and an understanding of the relevant business process. The measure chosen for the sickness and absence process was 'average days lost per year'. The target for this measure is 11.9 days lost per year per employee for sickness and absence. This benchmark was derived from the national average performance. The current performance was at 14.1 days lost per year per employee (Figure 5.6).

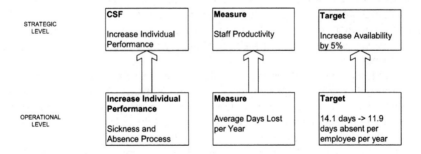

Figure 5.6. Deriving Operational Measures from Strategic Objectives

Discussion

The paper has described a methodological based approach to process improvement. In the setting of a police force the relative lack of competitive pressures driving innovation and the ideal opportunity to benchmark against other regional forces all undertaking similar tasks made a structured approach to process improvement (Klein, 1994) the most appropriate. In addition Buchanan (1998) finds that a "blank sheet" approach to organisational redesign leads to the construction of a change agenda which is broad and impractical, which conflicts with other organisational priorities (such as the implementation of change triggered by shifts in government policy), and which ignores significant barriers to change rooted in the sector's history and culture.

The first step in the approach is to derive a series of critical success factors for the organisational unit under scrutiny. In this case the HR division of a UK police force. The CSF's are placed within the framework of the balanced scorecard which provides a set of performance indicators for the stakeholders involved. Although a government organisation such as the Police is not in a competitive market it is necessary for it to conduct a detailed analysis to determine the requirements of what can be a diverse collection of stakeholders.

The next stage entails forming an understanding of the current process design. A process map presents a visual medium in which the process design can be discussed and agreed between interested parties. The use of a visual process map is seen as essential in providing personnel with an overall view of the process, in contrast to the particular aspects of the process they are involved with. Buchanan (1998) states that 'Process mapping, in whatever form, encourages a process orientation and overview'.

In order to focus the improvement effort a way of prioritising processes for improvement was required. Crowe and Rolfes (1998) state that little research has been undertaken in the subject of identifying process for improvement within a process improvement effort. They present a decision support system which evaluates business processes based on current performance and overall importance with respect to strategic objectives using influence diagramming software. In this case a spreadsheet-based scoring system was developed which prioritised processes in terms of their impact on meeting the CSF's derived in Step 1 of the methodology and the amount of potential innovation that was estimated could be gained from their redesign. The system provides a way of focusing on those processes which offer the most scope for improvement in the most relevant areas of performance.

The redesign stage utilises the areas defined in the ESIA framework (Peppard and Rowland, 1995) to redesign the processes identified. A variety of software tools are available to assist in redesign (Cheung and Bal, 1998). For instance a software tool that is increasingly being used to assist the redesign steps is Business Process Simulation which involves building a dynamic (time-based) animated model of the relevant business processes. Previous applications in the Police Service include the simulation of a custody suite (Greasley and Barlow, 1998). In this case the dynamic analysis offered by the simulation approach was not considered necessary as the timing of the sickness and absence events was not considered relevant to the redesign of their efficiency or effectiveness.

Finally a measurement system linking the operational business process performance with strategic targets was utilised. Government agencies such as the Police are often duplicated geographically across the country. With the publication of performance indicators, the use of benchmarking against 'best in class' performers can be utilised. In this case, as a first step in the improvement process, a target derived from a national average performance level was used as a benchmark.

Slack and Lewis (2002) state that the strategic management of any operation cannot be separated from how resources and processes are managed at a detailed and day-to-day level. The methodology described presents one way of ensuring the correct processes are identified and redesigned at an operational level in such a way as to support the organisation's strategic aims. In addition a performance measurement system is utilised to attempt to ensure that the changes implemented do actually achieve the desired effect over time.

Summary

This chapter has investigated the relationship between simulation and process improvement methodologies.

Case Study 1 demonstrates the use of Business Process Simulation (BPS) to assess business process performance as a result of the implementation of an information system for road traffic accident reporting. The information system offers a number of potential benefits including faster delivery of road traffic accident information to government agencies through the use of digital mapping technology. Faster process execution of document flows though workflow technology, rather than the movement of paper records, is also envisaged. Cost savings in terms of road traffic officer time are gained by eliminating the need to return to the police station to undertake administration duties. Savings are also predicted though a centralised administration facility, enabled through the digitalisation of data, rather than paper records kept at a divisional level. The quality of the process is improved by the greater accuracy of road traffic accident location analysis through the use of digital mapping and geocodes. Improved accuracy of RTA information will be gained though the use of a centralised database store replacing paper documentation. Although these potential improvements were recognised there was a lack of confidence in the introduction of information systems due to previous failures. The simulation was able to provide a level of confidence both in the operation of the new process design and quantification of potential benefits in terms of improved efficiency. Both of these factors provided an impetus to initiate the proposed changes.

Case Study 2 shows that by re-allocating tasks in custody of prisoner process as proposed provided a significant shift in resource utilisation away from the custody officer and on to the jailer. This re-allocation of tasks down the hierarchy could bring significant savings in staffing costs provided the staff time saved can be utilised effectively. Furthermore the results show that the jailer role can accommodate the re-allocated booking-in tasks, in addition to present duties, assuming the demand and process durations used in the model. The legal requirements surrounding the custody process constrained the amount of discretion available to personnel in the area and so facilitated the model building process. More loosely based activities would require the use of further experimentation to judge the effect on performance measures. It was envisaged that tasks within the custody area would evolve over time. A comprehensive menu system was implemented to enable users to reflect changes in the model. However certain changes, such as the introduction of new work processes, need the involvement of the analyst or training of end-users in the modelling system.

From the case study analysis the following main areas for the application of simulation can be identified in the context of process-centred change.

Ability to Measure Performance – Simulation allows the performance of both the present and reengineered systems to be assessed. Measurement of the present system may assist in identifying potential areas for improvement and serve as a benchmark with which to measure predicted performance of the reengineered process. The simulation is able to incorporate a range of performance measures, such as speed, cost and quality, which are determined by the process vision. In the case study the performance measures focused on the utilisation of staffing in order to minimise staffing costs. Another study could have focused on the speed of the custody booking-in process.

Ability to Try Alternatives – A key ability of simulation is its ability to quickly try alternative scenarios without disrupting the business system. The user is able to observe the sensitivity of system performance to changes in levers derived from the process vision. In the case study the simulation model proved vital in establishing performance under different staffing scenarios. In particular the nature of the demand, with a number of arrests occasionally occurring simultaneously, put particular demands on the system which were hard to predict. In this case actual arrest data was used to simulate demand in order to validate and maximise confidence in the model.

Ability to Communicate Process – Reengineering involves making radical change to business processes. Simulation can provide powerful assistance in helping to generate new change ideas, explore the effect of changes and bring those changes about. A full-screen animation of a system both before and after a proposed change provides a powerful tool for understanding and therefore lessons the risk of implementation. Understanding is gained through observing the interactions between elements in the system over time. The ability to 'fast forward' the simulation through time helps understanding of relationships which would not normally be apparent because of the time delays involved between the cause and effect cycle.

Case Study 3 describes a structured approach to process improvement in the context of the HR division of a UK police force. The approach has combined a number of established techniques of process improvement such as the balanced scorecard and process mapping with a scoring system developed to prioritise processes for improvement. The case demonstrates the need to choose and in some cases develop in-house tools and techniques dependent on the context of the process improvement effort. In this case the context was defined by the role of the Police as a public sector agency with a number of stakeholders and its emphasis on the management and deployment of human resources.

References

Aguilar, M.; Rautert, T.; Pater, A. (1999), Business Process Simulation: A fundamental step supporting process centred management, *Proceedings of the 1999 Winter Simulation Conference,* ed. by Farrington PA, Nembhard HB, Sturrock DT, Evans GW, SCS, 1383–1392.

Bell, P.C.; Anderson, C.K.; Staples, D.S.; Elder, M. (1999), Decision-makers' perceptions of the value and impact of visual interactive modelling, *Omega*, 27: 2, 155.

Buchanan, D. (1998), Representing process: the contribution of a reengineering frame, *International Journal of Operations and Production Management*, 18: 12, 1163–1188.
Campbell, S. and Kleiner, B.H. (1997), New developments in re-engineering organizations, *Work Study*, 46: 3, 99–103.

Cheung, Y. and Bal, J. (1998), Process analysis techniques and tools for business improvements, *Business Process Management Journal*, 4: 4, 274–290.

Cotoia, M. and Johnson, S. (2001), Applying the axiomatic approach to business process redesign, *Business Process Management Journal*, 7: 4, 304–322.

Crowe, T.J. and Rolfes, J.D. (1998), Selecting BPR projects based on strategic objectives, *Business Process Management Journal*, 4: 2, 114–136.

Doyle, P. (1998), *Marketing Management and Strategy*, 2nd Edition, Prentice Hall.

Fathee, M.M.; Redd, R.; Gorgas, D.; Modarres, B. (1998), The Effects of Complexity on Business Process Reengineering: Values and Limitations of Modeling and Simulation Techniques, *Proceedings of the 1998 Winter Simulation Conference*, ed. by Mederios DF, Watson EF, Carson JS, Manivannan MS, SCS, 1339–1345.

Giaglis, G.M. (1999), On the Integrated Design and Evaluation of Business Processes and Information Systems, *Communications of the AIS*, 2.

Giaglis, G.M.; Mylonopoulos, N.; Doukidis, G.I. (1999), The ISSUE methodology for quantifying benefits from information systems, *Logistics Information Management*, 2, 50–62.

Greasley, A. (2001), Costing Police Custody Operations, *Policing: An International Journal of Police Strategies & Management*, 24: 2, 216–227.

Greasley, A. and Barlow, S. (1998), Using simulation modelling for BPR: resource allocation in a police custody process, *International Journal of Operations and Production Management*, 18: 9/10, 978–988.

Gulledge, T.R. and Sommer, R.A. (2002), Business process management: public sector implications, *Business Process Management Journal*, 8, 364–376.

Harrington, H.J. (1991), *Business Process Improvement*, McGraw-Hill, New York.

Harrington, H.J. and Tumay, K. (2000), *Simulation Modeling Methods: To reduce risks and increase performance*, McGraw-Hill.

Harrison, A. (1998), Investigating business processes: does process simplification always work?, *Business Process Management Journal*, 4: 2, 137–153.

Johansson, H.; McHugh, P.; Pendlebury, J.; Wheeler, W. (1993), *Business Process Re-engineering: Break Point Strategies for Market Dominance*, John Wiley & Sons, Chicester.

Kaplan, R.S. and Norton, D.P. (1996), *The Balanced Scorecard: Translating Strategy into Action*, Harvard Business School Press, Boston.

Kelton, W.D., Sadowski, R.P., Sadowski, D.A. (1998), *Simulation with Arena*, McGraw-Hill, Singapore.

Kelton, W.D.; Sadowski, R.P.; Sadowski, D.A. (2001), *Simulation with ARENA*, McGraw-Hill, Singapore.

Kettinger, W.; Teng, J.; Guha, S. (1997), Business process change: a study of methodologies, techniques and tools, *MIS Quarterly*, March, 55–80.

Klein, M. (1994), Re-engineering methodologies and tools: a prescription for enhancing success, *Information Systems Management*, Spring, 30-5.

Levas, A.; Boyd, S.; Jain, P.; Tulskie, W.A. (1995), Panel discussion on the role of modelling and simulation in business process reengineering, *Proceedings of the 1995 Winter Simulation Conference*, ed. by C. Alexopoulos, K. Kang, W.R. Lilegdon and D. Goldsman, SCS, 1341–1346.

Levine, L.O. and Aurand, S.S. (1994), Evaluating Automated Work-Flow Systems for Administrative Processes, *Interfaces*, 24, 141–151.

Majed, A. and Zairi, M. (2000), Revisiting BPR: a holistic review of practice and development, *Business Process Management Journal*, 6: 1, 10–42.

Martilla, J.A. and James, J.C. (1977), Importance-Performance Analysis, *Journal of Marketing*, January.

McAdam, R. and Bailie, B. (2002), Business performance measures and alignment impact on strategy: The role of business improvement models, *International Journal of Operations and Production Management*, 22: 9, 972–996.

Neely, A.; Gregory, M.; Platts, K. (1995), Performance measurement system design: A literature review and research agenda, *International Journal of Operations and Production Management*, 15: 4, 80–116.

Painter, M.K.; Fernandes, R.; Padmanaban, N.; Mayer, R.J. (1996), A methodology for integrating business process and infrastructure models, *Proceedings of the 1996 Winter Simulation Conference*, ed. by Charnes JM, Morrice DJ, Brunner DT, Swain JJ, SCS, 1305–1312.

Peppard, J. and Rowland, P. (1995), *The Essence of Business Process Re-engineering*, Prentice Hall.

Petrozzo, D.P. and Stepper, J.C. (1994), *Successful Reengineering*, Van Nostrand Reinhold, New York.

Pidd, M. (1998), *Computer Simulation in Management Science*, 4th Edition, John Wiley & Sons Ltd, Chichester.

Profozich, D. (1998), *Managing Change with Business Process Simulation*, Prentice Hall, New Jersey.

Radcliffe, J. (2000), Implementing and integrating crime mapping into a police intelligence environment, *International Journal of Police Science and Management*, 2, 313–323.

Robinson, S. (1994a), *Successful Simulation: A Practical Approach to Simulation Projects*, McGraw-Hill, Berkshire.

Robinson, S. (1994b), Simulation projects: building the right conceptual model, *Industrial Engineering*, September, 34–36.

Robinson, S. (2001), Soft with a hard centre: discrete-event simulation in facilitation, *Journal of the Operational Research Society*, 52, 905–915.

Sheehan, M. (2000), Police chiefs crack down on shirkers, *The Sunday Times*, February 27, 9.

Slack, N. and Lewis, M. (2002), *Operations Strategy*, Pearson Education Limited, Harlow.

Tumay, K. (1996), Business Process Simulation, *Proceedings of the 1996 Winter Simulation Conference*, ed. by Charnes JM, Morrice DJ, Brunner DT, Swain JJ, SCS, 93–98.

Verma, R.; Gibbs, G.D.; Gilgan, R.J. (2000), Redesigning check-processing operations using animated computer simulation, *Business Process Management Journal*, 6, 54–64.

Warren, J.R. and Crosslin, R.L. (1995), Simulation Modeling for BPR, *Information Systems Management*, 12, 32–43.

Wiley, R.B. and Keyser, T.K. (1998), Discrete event simulation experiments and geographic information systems in congestion management planning, *Proceedings of the 1998 Winter Simulation Conference*, ed. by Mederios DF, Watson EF, Carson JS, Manivannan MS, SCS, 1087–1093.

Yu, B. and Wright, D.T. (1997), Software tools supporting business process analysis and modelling, *Business Process Management Journal*, 3: 2, 133–15.

6

Enabling Simulation – Qualitative Simulation

Introduction

Simulation is associated with providing an outcome in terms of quantitative performance statistics for measures of interest such as machine utilisation rates and process cost estimates. However it has been found that there are also useful qualitative outcomes to a simulation study that add to its usefulness and provide a further case for its more widespread usage. This chapter investigates on the qualitative outcomes of three simulation studies.

Case Study 4 shows how simulation animation facilities are particularly useful in communicating the operation of dynamic and interacting systems such as transportation facilities. In this case study a novel use of simulation was to show capability to a third-party client in meeting service level targets for the operation of a train maintenance depot.

Case Study 5 reports that an assessment of the use of simulation should incorporate factors such as the potential of using the model building process to gain understanding of a system and the use of animation to communicate ideas.

In Case Study 6 a discrete event simulation model was developed and used to estimate the storage area required for a proposed overseas textile manufacturing facility. It was found that the simulation was able to achieve this because of its ability to both store attribute values and to show queuing levels at an individual product level. It was also found that the process of undertaking the simulation project initiated useful discussions regarding the operation of the facility.

Case Study 4
Using Simulation Modelling to Assess Service Reliability

Introduction

In service operations the product is a process, i.e. a method of doing things. Thus the design of the service is the specification of how the service should be delivered

(Evans, 1990). Schmenner (1993) describes three major components of a service system. The *service task* defines what the operation must do well to satisfy customer needs. This is translated into a set of *service standards or levels* which define the tasks and constraints surrounding the system. The *service delivery system* supports the service task and maintains the designated service levels, which provide a measure of quality. Lovelock (1988) provides a definition of service quality as 'Quality is the degree of excellence intended, and the control of variability in achieving that excellence, in meeting the customers requirements.' The definition reflects the fact that there can be a large variation in the demand level on the service delivery system. Maintaining service quality levels during these variations is a major task of service quality management. Fitzsimmons (1994) describes a number of 'tools of TQM' such as flow charts and Statistical Process Control (SPC) which can ensure that a process can achieve quality of conformance (i.e. provide a consistently quality service). This paper describes the use of the simulation technique to ensure that a service delivery system can undertake a number of service tasks to specified service levels.

Case Study

The study concerns a major UK based manufacturer of railway rolling stock and equipment. The company was preparing a bid for a major order to construct a number of rail carriages for an underground transportation system. The bid included a service contract to maintain the carriages over their lifetime. This would entail operating a maintenance depot on existing track and rail infrastructure.

Although the company had extensive knowledge in rail vehicle design and construction it had less experience of maintenance operations. However to secure the contract it would be necessary to show capability in meeting service quality levels for the depot. These quality indicators were contained in a service-level agreement and focused on ensuring the reliability of service supply, i.e. the client had to demonstrate that refurbished trains would be available, at a time specified by the train timetable, every time. Penalty fines would be imposed if service levels did not match the targets set in the agreement.

In this case the requirement to operate the depot on existing infrastructure meant that capacity would be constrained by the number of existing stabling and refurbishment lines. Demand was also fixed in terms of the quantity of trains requiring refurbishment and the arrival time at the depot which was derived from the local train timetable. Thus management needed to assess if sufficient capacity was available in order to carry out the refurbishment tasks in the time period between delivery to the depot and request for next service.

After a feasibility study it was decided to use a simulation model to address both the operational issues involved in running the depot and to use as a tool to assist in proving capability of operating the depot to the client.

The depot is located on a current site with a fixed number of lines and facilities available. Carriages are delivered from the main underground line to a changeover point to the left of the depot site. The entry and leave time for carriages to the depot are determined by the local train timetable and so is not under the control of the client. A single shunter is utilised to transfer carriages both to and from the depot and for movement within the depot site. On entry carriages move along the sidings and progress to the stabling points, workshop, cleaning shed or lift shop facilities as appropriate. The depot consists of 14 stabling points for carriages that are either waiting for service or maintenance. In addition there is a covered workshop consisting of four available lines, a cleaning shed of 2 lines and a lift shop which can accommodate a further carriage.

The demand on the depot consists of scheduled trains who arrive for stabling and refurbishment and 'casualties' which are trains that have developed a fault while in operation and return to the depot for repair.

Scheduled Trains
The existing timetable operates with 23 carriages entering the depot a day at a time pre-defined from the train service timetable. The carriages are then submitted for one of three operations. A 14-day clean lasts on average 2 hours and occurs once a fortnight. A more extensive heavy clean lasts on average 4 hours and takes place in the cleaning shed. Carriages not due for cleaning are stored at stabling points until needed. 3 lines are available in the workshop for the 14-day clean operation and 2 lines are available in the cleaning shed for a heavy clean. The trains are rotated between the maintenance depot and other depots on the underground system to ensure all trains in the sector are cleaned as appropriate.

Casualties
In addition to processing scheduled trains the depot must have the capability of repairing trains which develop a fault during an operation with the minimum of disruption to service delivery. It is assumed that a train which develops a fault (called a casualty) can be returned to the depot. Thus casualties arrive at a predefined rate to the depot and are replaced by a train ready for service. If no trains are ready for service the casualty is replaced as soon as one becomes available. The casualties enter the depot and can make up to 5 visits to the workshop (lines 1 to 4) for assessment and repairs. Each visit is separated by a wait at a stabling point for a predefined duration to simulate an interruption to the repair process caused by factors such as a wait for parts. Once the train has been repaired it is now ready for service.

The Simulation Study

The study was structured in two stages. Stage one involved building a model of the depot without casualty arrivals. This allowed the project team to concentrate on the task of forming rules which govern the movement of the carriages within the depot by the single shunter. Because of the limited line capacity available and the need to ensure a clear path for movements a clear set of operational rules for train movement was seen as critical in maximising performance. Once a feasible model was operational then the second stage of the project involved inserting randomly generated casualty arrivals into the model and observing the effect on service level performance. This approach is recommended by Hill (1983) who states that in the provision of most services, demand can be separated into planned and random and analysed separately in order to smooth the overall demand pattern.

The methodology for the simulation study is based on that of Pegden (1995) and is shown in Figure 6.1.

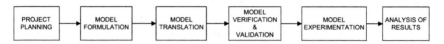

Figure 6.1. The simulation study methodology

Project Planning

As a first step a project team was formed consisting of the author as consultant to construct the simulation, an Industrial Engineer based at the company with experience of simulation, the project manager for the proposal and a project engineer who would be responsible for depot operations. A variety of data was collected including a local train timetable from which depot arrival and departure times could be estimated, timings for depot operations and a depot layout plan.

Model Formulation

In order to provide an initial analysis of the problem the author suggested using a Gantt chart (Stevenson, 1993) to map out the train arrivals and departures. Following meetings of the project team a chart was produced which showed the arrival and departure of trains during the day and from this a subsequent capacity loading profile could be derived for the stabling, workshop and cleaning shed

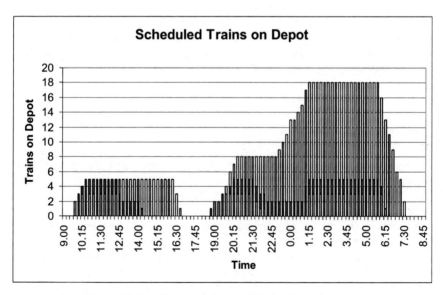

Figure 6.2. Loading graph for scheduled trains at the depot

resources. (Figure 6.2). It was clear from the loading graphs that the focus of the study would involve ensuring that a service would be maintained despite the changes in demand caused by the peak morning and afternoon service. As trains begin leaving the depot at peak service time, the depot is near capacity and so too many casualties entering the depot at this time could overload the facilities. Also when no trains are on depot and all are out to service, casualties arriving cannot be replaced immediately, causing loss of service.

The next step was to formulate rules for the movement of carriages within the depot. This was achieved by mapping out the positions of carriages in the depot after each carriage movement on an enlarged schematic of the depot. After a number of walkthroughs a series of rules were generated. A conceptual model was developed which outlines the main movements of carriages within the depot to provide a guide for the model translation stage (Figure 6.3).

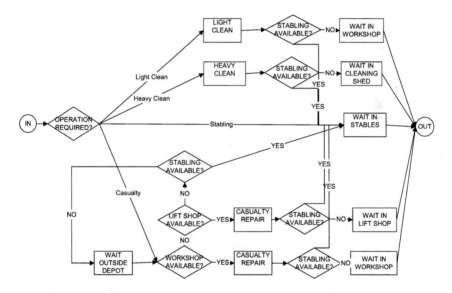

Figure 6.3. The maintenance depot conceptual model

Model Translation
A model was generated using the SIMAN/CINEMA™ simulation system. This contains a number of constructs relevant for transportation problems and provides excellent animation facilities (Pegden, 1995). SIMAN is a computer software language designed for constructing simulation models, which can be run and generate results independently of the CINEMA facility. CINEMA is a computer draw package that allows background and animated objects to be specified. A screen display of the depot simulation is shown in Figure 6.4.

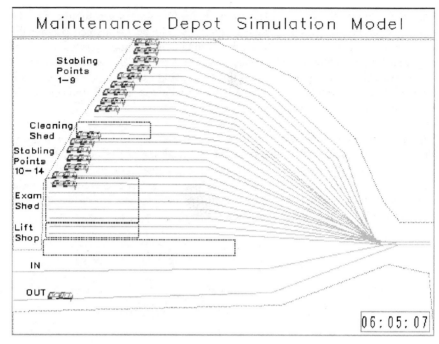

Figure 6.4. Depot Simulation Model Display

The display is in the form of an overview schematic which shows the logic governing the movement of carriages within the depot.

Model Validation and Verification
Verification is the process of checking that the simulation coding provides a current logical representation of the model. This process includes aspects such as ensuring clarity of coding design, extensive debugging and adequate documentation. Validation is the process of checking that the model provides an adequate representation of the real-world system. An important aspect of validation is that the client of the model has confidence in the model results. In this case a major objective of the simulation was to model the movement of carriages within the depot. This was achieved by using the animation facilities of the simulation software. Although the disadvantages of using animation to validate a simulation model have been noted (Law and McComas 1989; Paul 1991), in this case it proved critical in enabling the operation of the facility to become readily observable and describable, thus increasing the level of confidence regarding how closely the simulation represented the problem itself (Rowe and Boulgarides 1992). A number of walkthroughs were conducted with the project team which led to a number of refinements to the rules governing carriage movements in order to generate more realistic behaviour. For example, carriages processed shortly before peak service time would remain in the workshop or cleaning shed as appropriate, thus eliminating unnecessary shunting movements to the stabling lines.

When a satisfactory operation was achieved the fixed process durations could be replaced by probability distributions. The maintenance processes (i.e. light clean in the workshop and heavy clean in the cleaning shed) were modelled using a triangular distribution with low and high values 10% from the mode. The arrival time between casualties was modelled using an exponential distribution with a mean derived from the casualty rate per day. Data was also supplied on the process duration of visits required to the workshop to repair faults and these were modelled in Table 6.1.

Table 6.1. Fault Repair time

Percentage of casualty arrivals	Mean Process Duration (minutes)
66	120
23	180
5	300
6	420

The process duration was modelled as a triangular distribution with the minimum and maximum values 10% from the mode. Once the behaviour of the casualty trains was validated experimentation on the full model could begin.

Model Experimentation
Due to the random nature of the distributions used in the model the animation facilities were used to observe the performance of the model over a number of simulation runs. It became clear that service levels could not be maintained with the casualty rate envisaged and that new operating procedures had to be devised. The performance of the depot was observed using different operating procedures over a number of runs until satisfactory operation was ensured. Hope and Mühlemann (1997) outline two main approaches to capacity management in services. One approach is the 'level capacity' strategy in which capacity levels are seen as fixed. In this case an attempt is made to influence demand to meet capacity. The alternative approach is the 'capacity following demand' strategy when capacity levels are altered to meet demand levels by adjusting resource levels. In this study the model incorporated a mix of the two approaches in order to ensure reliable service delivery. The way these two approaches were put into operation in this study is outlined below.

In order to avoid overloading the depot at peak demand it was decided to only allow repairs to casualty trains at off-peak times, represented in the model by a workshop 'window'. This allowed the user of the model to specify time slots in which casualty trains could not enter the workshop but would remain at a stabling point. Thus at peak time only the workshop line 4 is available for casualties while at off-peak times lines 1 to 4 are available.

The assessment of the capacity of the depot to service scheduled and casualty trains was used to assess resource requirements for a preventative maintenance programme.

If depot capacity is full a queue of trains will form at the depot entrance. The simulation is able to provide statistics on queue length and time in queue for trains under a number of scenarios.

Another problem was the absence of trains on depot after they had departed for peak service and before any had returned. To ensure any casualty trains could be replaced immediately at this time two 'spare' trains, over and above those required to meet service levels, were placed on depot to replace casualties.

A line in the Lift Shop is available to casualties for emergency use only. (I.e. if there is no other space for the casualty to be repaired or stored). Again this strategy is designed to ensure service levels are met.

The simulation experimentation was designed to both prove that the depot could operate without 'missed service' events (a missed service is a request for a train from the timetable that is not met immediately) during normal service with no casualty events occurring. The second phase of the experimentation was to assess the performance of the depot with a range of casualty arrival rates. The effect of placing one or two 'spare' trains on depot on performance was also investigated.

Results

The results of the simulation experimentation are shown in Table 6.2.

Table 6.2. 'Missed Service' events by casualty rate and spare trains

Casualty Rate per Day	'Spare' Trains on Depot	Mean 'Missed Service' Events in 4-Week Period
0	0	0
1	0	1.4
	1	0.4
	2	0
2	0	2.6
	1	1.6
	2	0.6
3	0	2.8
	1	1.8
	2	0.8

The simulation was run for a time period of 28 days for each of 10 scenarios. The time period of the simulation run was chosen to provide sufficient time for a number of 'missed service' events to occur, whilst not requiring excessive computer time to record. Results were collected of missed service events for 5 simulation runs. The mean value of these observations was computed to provide a measure of the average number of missed service events that could be expected over the four week time period of the simulation run. The simulation was configured with starting conditions to minimise initial bias and the method of independent replications was used to analyse the results of the simulation model. Thus a different random number stream was used for each simulation run. Five runs were performed to ensure that the effect of random variation (in casualty arrival times and maintenance process times) could be assessed. The accuracy of the results could be improved with additional simulation runs.

The results show that with no casualty arrivals and no spare trains, no missed service events are likely to occur. With a casualty rate of 1 per day and 2 per day, two spare trains are required to ensure no missed service events. With a casualty rate of 3 casualties per day, even two spare trains will give an 80% chance of a missed service event during a 4-week operating period. This performance could be improved by reducing the failure rate of trains, through preventative maintenance (Bentley, 1999) for example, or increasing depot capacity through decreasing maintenance and repair process times or increasing the number of spare trains available.

Discussion

The simulation was successful in showing how the depot could operate to the service levels specified through the collection of data on the number of 'missed service' events. A number of changes to the basic depot design were assessed and the ability to meet 'normal' service demand proved.

The simulation was then able to assess the performance of the depot with a randomly distributed arrival of 'casualty' trains that represented an extra demand over the normal maintenance cycle. It would be difficult to assess the performance of the depot by other means because of the variation in maintenance process and repair times and casualty arrival rates. The timing of casualty arrivals may also lead to varying performance as the demand on the depot is not constant during the day. The simulation model was able to assess the effect of introducing 'spare' trains on depot to buffer the system from these demand variations.

A further development occurred when the project engineer became concerned that future changes in timetable requirements would significantly alter the demand requirements on the depot. Thus he requested that the model be adapted to allow him to carry out further investigations alone. This was achieved by the use of a menu system that allows the user to change the arrival and leave times for the trains. The need for training in statistical techniques to analyse the simulation output was then addressed. In this case the measurement of a discrete variable ('missed service') led to a relatively simple statistical analysis. In other cases however the use of a t-test or other method to ascertain statistically significant results would be required which may necessitate training of users in these techniques (Hollocks, 1995).

Case Study 5
The Case for the Organisational Use of Simulation

Introduction

The study concerns an autonomous division of a major UK based manufacturer of railway rolling stock and equipment. The plant manufactures a range of bogies which are the supporting frame and wheel sets for rail vehicles. The company has a history of supplying the passenger train market in the UK but over a period of time low demand and increased competition had led it to enter new markets including European inner-city transport and the supply of freight bogies to Far East countries. The need to compete on a global basis led the company to re-evaluate its manufacturing facility with particular emphasis on the need to increase output, reduce lead times and increase flexibility. To meet these demands management had identified areas where substantial investment was required.

Case Study

The facility layout is on a line basis with the manufacturing process consisting of six main stages of fabrication, welding, frame machining, paint, fitting and quality audit. Each stage must be completed in order before the next stage can begin. The stages are now briefly described:

Fabrication
The fabrication stage prepares the bogie frame sections from sheet steel and bought-in castings. A custom template is designed from which the parts required are cut from sheet steel to standard batch sizes. Parts not needed immediately are held in storage. Processed parts and castings are brought together to form a bogie 'kit' which is assembled on a jig and taken to the subsequent welding stage.

Welding
A bogie sub-assembly is manually welded on a jig at a work station to form a main bogie frame.

Frame Machining
The main bogie frame is then transferred to a C.N.C. centre for the machining of any holes or bores needed for the fixing of sub-assemblies such as the braking and suspension systems. Bogies are fixed to a slave table and the machine processes the frame according to a preset operation sequence.

Paint
The frame is then manually painted while being suspended from an overhead moving circular track.

Fitting
Manufactured sub-assemblies and bought-in components such as motors are then assembled on the bogie frame. The frames are placed on supports and are moved along a line at different stages of assembly with overhead cranes.

Quality Audit
Final inspection is carried out to ensure all bogies meet the required specification. It was usual that a certain amount of paint touch-up work is required at this stage due to damage caused to the paint finish during the fitting stage.

The Simulation Study

The focus of the study was on product layout design with the main objective being to ensure that the performance of the whole manufacturing system would meet required output levels. The output level was converted into a target cycle time (i.e. time between manufacture of products or output rate). As stated the product layout consists of six main stages with the product passing through each stage in turn. This means that the effective cycle time for the whole system is determined by the stage with the longest cycle time. Thus the simulation study was to ensure that any investment in a particular production stage would not be nullified by a longer cycle time elsewhere. The simulation would provide an analysis tool to signal any process improvement activities needed before installation took place. The objective was to obtain a balanced line (i.e. all cycle times equal) which would enable a smooth parts flow through the production stages. This would facilitate the introduction of a pull-type Just-In-Time (J.I.T.) production control system (Schniederjans, 1993) to replace the present push Materials Requirement Planning (M.R.P.) make-to-stock system.

The project team for the study included the Production Manager, an Industrial Engineer, an internal consultant (from another site) with some experience of simulation and the author acting as an external consultant. In the event when the objectives had been agreed most contact was made between the author and the Industrial Engineer based at the facility. Information was supplied by the Industrial Engineer on components flows, setup and process times and other relevant information. Most of the data was gathered from the M.R.P. control system in use. The model was built using the SIMAN/CINEMA™ system (Pegden et al., 1995). This provides extensive manufacturing modelling facilities as well as a high resolution graphical display. A plan view of the manufacturing facility was constructed with an animation of parts moving through the system. Figure 6.5 shows a 'zoomed' portion of the simulation model display.

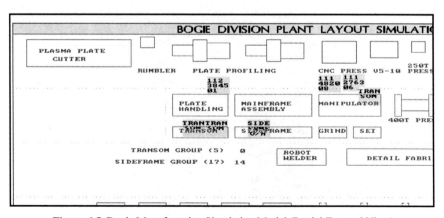

Figure 6.5. Bogie Manufacturing Simulation Model (Partial Zoomed View)

The debugging and animation facilities of SIMAN/CINEMA™ were used to ensure that the model was free from any errors. The model was validated by simulating the present system and discussing the results with management to identify any anomalies.

Simulation Model Analysis

Line Balancing
The first stage of the simulation analysis was driven by the need to reduce cycle time (i.e. maximise production output) below a target level. Any stage above the target was investigated by the project team and changes made to machinery or working practices suggested. The effect of these changes was entered into the model and the simulation was re-run to observe the effect on whole system performance. To achieve a synchronised pull system it was necessary for the fabrication process to only produce material when needed. This meant operating with a batch size of one bogie. At present cutting templates were individually designed to produce batches of parts when needed. They were replaced with a standard template which produced only the parts for one bogie and proceeded through the fabrication stage as before. From welding times provided for the bogie assemblies a welding line configuration was constructed. Manual welding was used and welding times were considered fixed so to meet demand additional welding lines were utilised. Figure 6.6 shows the observed average cycle times from the simulation for each production stage and the major fabrication processes. The graph shows clearly where management effort needed to be directed to achieve the target cycle time.

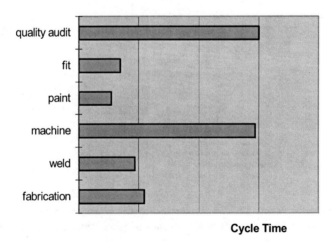

Figure 6.6. Present System Cycle Time Performance

The quality audit stage was set at a nominal amount by management. Significant problems had occurred at this stage with the spray finish on the bogie frames being damaged during the sub-assembly fitting stage. This had to be rectified by a manual touch-up process which could take longer than the original spray time. The paint area would also need to be re-configured due to new environmental controls. The problems had been recognised by management and an investment in an epoxy paint plant producing a hard wearing finish was planned.

The bogie frame machining centre had previously been recognised by management as a bottleneck process. The bogie frame went through a number of pre-programmed steps on the machine and the cycle time was dependent on the capability of the machining centre itself. Consequently a major part of the planned investment was a new machining centre with a quoted cycle time below the target. An investigation of the fabrication processes revealed that although the cycle times were above target, the majority of this time was used for machine setup. Figure 6.7 shows the effect on cycle time of a reduction in setup time of 10% to 90%.

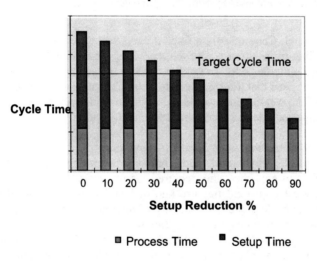

Figure 6.7. Setup Reduction for Fabrication Stage

From Figure 6.7 it is clear that to achieve the target cycle time a setup reduction of 50% is required. A team was assembled to achieve this target and it was met by the use of magnetic tables to hold parts ready for processing. The simulation was run with the target demand cycle time to assess system performance. Surprisingly a queue of parts was observed in front of the Edge Bevel machines indicating a bottleneck process which could not achieve the target cycle. Further investigation revealed that the sequence of parts introduced on to the two machines led to an uneven distribution of loading on one machine taking it

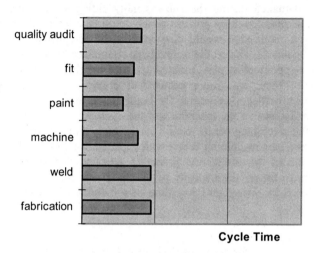

Figure 6.8. Proposed System Cycle Time Performance

over the target hour limit. Consequently a fixed process sequence was devised to ensure an even loading distribution across the machines. The simulation was re-run and the results (Figure 6.8) show the system achieving the required performance. It can be seen that a further reduction in fabrication setup times and a re-configuration of the welding line would reduce the overall cycle time further, producing a more balanced line and increasing capacity utilisation.

Assessing the Effect of Disturbances on System Performance
The second stage of the simulation study involved assessing the effect of disturbances on the system once the required performance had been met. Management was interested in simulating breakdown events on the Edge bevelling machines which had no backup and had a history of intermittent breakdown. The experiment was based on an average machine failure of 3 times a working week with an average repair time of 40 minutes. It was assumed breakdown events were dependent on time, not the number of pieces processed on the machine. The high utilisation (over 90%) made these two approaches essentially the same. An exponential distribution for time between breakdowns and a 2-Erlang distribution for breakdown length was used. These distributions were chosen using a combination of observation of graphs of historical data and reference to theoretically appropriate distributions (see Law and Kelton, 2000). It was found that with random breakdowns representing a 3% loss of capacity the system was still able to meet the target cycle time.

Investigation of the Facility Layout
With the proposed new investment in machinery and the desire to achieve maximum performance the simulation was used to investigate various facility

layouts within the factory. The simulation was able to show a clear picture of the flow of parts over time, although as no routing times were changed the simulation results were identical for different runs. Adjustment to the layout is a time consuming process and needs in-depth knowledge of the simulation package used. Consequently its usefulness is limited in allowing management to experiment with various layouts.

Simulation Study Results

By implementing the changes outlined in the study the simulation was able to predict the following improvements in performance (Table 6.3).

Table 6.3. Results Table

Performance Measure	Change (%)
Cycle Time	−65
Lead Time	−19
Output per Week	+220
Cycle efficiency*	+29

*Cycle Efficiency = 100%- % Idle Time where %Idle Time = Idle time per cycle / total cycle time

These are substantial improvements in performance and meet the output targets set by management. However the simulation results shown in Figure 6.8 show that further reductions in the cycle times for the fabrication and weld stages would lead to a further increase in cycle efficiency, reflecting a more balanced line, and thus a further increase in output (Shin and Min, 1991).

Discussion

It is clear from the case that the simulation model achieved its aim in terms of providing a tool for management in improving system operation before the introduction of the proposed machinery. This highlights the main advantage of the simulation method: the reduction of risk. 'Uncertainty is removed and replaced with certainty about the expected operation of a new system or about the effects of changes to an existing system' (McHaney, 1991). The simulation was useful because of its ability to provide the 'whole' picture of the process and demonstrate the frailty of local solutions (Robinson, 1994).

However within the case company there was a lack of awareness of the potential benefits of simulation which are discussed below under the heading 'organisational issues'. It was also noted that the extent of the use of simulation will depend on the usability of the software tool itself and this is discussed under the heading 'simulation software issues'.

Use as a 'One-Off' Technique

The research found that there was still a perception within the Industrial Engineers and management involved in the project that the simulation study was a 'one-off' project. Once the simulation was used and the decision made regarding the introduction of machinery the simulation model was put to one side. On questioning whether there would be any further use for the model one manager said that 'the simulation had done its job and paid for itself'. There was no attempt by any of the project group to see if the simulation model could be extended and used in another area, or as part of the training and education process. It was seen as a decision making tool and once the decision was made then it was redundant.

Expertise Transfer

The above lack of awareness of the scope of the technique was found to be connected to other findings within the case study. In particular the lack of technical skills and understanding in applying and interpreting the results of simulation. The author acted as a consultant on the model building stage of the project and was surprised at the lack of understanding of the technique. Only the internal consultant had some knowledge but he was situated on a different site. The project group seemed unwilling to try and learn how simulations worked and instead of training and educating their own staff with the necessary skills which could be drawn upon at any time, they preferred the use of consultants. This has a number of effects in not only achieving a superficial understanding of the simulation process but a lack of appreciation of the potential for simulation in other areas. Without understanding it they will not be able to adopt it, for example in the training area of the change process. This is combined with the observation that even those who do know how to use simulation become "experts" within a technically oriented environment. This means that those running the business do not fully understand the technique which could impact on their decision to use it or go with the decisions. Thus the point of using simulation as a decision making tool for managers is becoming lost within the technical computerised sphere of a small number of individuals. Although within the case study there were no "experts" the use of the simulation was still restricted to a small number of people.

Communication Tool

A further finding from the case study was the lack of involvement from the shop floor in the development of the simulation model and the subsequent decision. The company were going through a period of organisational change, not only in the way they manufactured the product but also in their approach towards human resources. They were attempting to move away from adversarial labour relations with a multi-union site to a single union agreement, and a reorganisation of the management hierarchy to a much flatter system with managers reporting directly to the Managing Director. Within this context of change the company could have used the simulation project to develop involvement from the shop floor in a number of areas. When requiring information on the current configuration of machinery the Industrial engineers and management were not inclined to ask the shop floor. If they were unsure a guess was made between the groups rather than approach those on the shop floor. Although this probably reflects the traditional nature of the organisation it was an ideal opportunity in the current climate of change to break the traditional mould and move to shop floor input on the project. Simulation can be a strong facilitator of communicating ideas up and down an organisation. Engineers for example could have used the simulation to communicate the reasons for taking certain decisions to shop-floor personnel who might suggest improvements. Kaizen and continuous improvement programs within companies has provided benefits in terms of quality, productivity and efficiency (Imai, 1986; Monden, 1983; Bessant et al., 1994). The use of simulation as a tool for employee involvement in the improvement process could be a vital part of the overall change strategy.

Simulation Software Issues
One approach for assessing the capabilities of simulation software for on-going decision support is to regard it as a Decision Support System (DSS) whose components can be categorised as dialog (interface facilities), data (database integration) and modelling (constructs) (Sprague, 1993). The case system was modelled using the SIMAN/CINEMA™ package which provides extensive manufacturing modelling capabilities but requires programming expertise and does not incorporate an integrated menu system or database facilities. Its replacement is the ARENA™ system (Kelton et al., 2007) which will be used as an example of a Visual Interactive Modelling (VIM) tool (Pidd, 1998) and assessed in terms of its dialog, data and modelling capabilities.

In term of the dialog or user interface component the ARENA system incorporates the Microsoft VBA programming language. This provides the capability to custom design user input menus and output results screens using the standard Visual Basic approach of forms and event coding. All the ARENA elements in terms of model variables and statistical information are available to the VBA code. In addition access is available to other VBA enabled products such as the EXCEL spreadsheet and the VISIO drawing package. This allows applications such as EXCEL to provide statistical and graphical analysis (Seppanen, 2000). Although the ability to design user friendly input and output displays enables easier use of the simulation by end users, they do require knowledge of the Visual Basic language to be implemented.

In terms of the database integration ARENA can import and export information from Microsoft Access and Excel packages. Harrell (Diamond et al., 2002) outlines the need for simulation to integrate more seamlessly with enterprise databases and other enterprise applications such as ERP systems and supply-chain analysis tools.

In terms of modelling capabilities models are built in the ARENA system by placing icons representing elements such as processes and decisions on the computer screen. These icons are connected to represent the path of entities (representing items such as products, people and information) through the process. The model is then run over a time period and statistical information is provided on aspects such as resource utilisation and queue times. A variety of developments in the modelling capabilities of visual interactive modelling systems are taking place. For example (Thomas and Dessouky, 2003) outline the use of the Microsoft VISIO drawing system as a mainstream software package that can be used to enter simulation logic details in place of the simulation environment. Pegden (Diamond et al., 2002) puts forward the use of a library of model components that can be used for many applications as a way of simplifying model development. Ingalls (Kachitvichyanukul et al., 2001) notes that simulation environments are designed for specialist simulation developers. A key growth area for simulation may be to embed a simulation capability into a larger system. An environment such as Microsoft Visual Studio, where the simulation would look like any other embedded system, is suggested as more appropriate for this type of use. Also Schmeiser (Kachitvichyanukul et al. 2001) notes the lack of sophisticated statistical analysis support provided with most simulation software, while acknowledging the main reason for this could be the lack of statistical knowledge of model developers. Hollocks (1995) indicates the need to train users, in addition to developers, in statistical techniques to ensure correct interpretation of model results.

Case Study 6
Using Simulation for Facility Design

Introduction

A discrete event simulation model was developed and used to estimate the storage area required for a proposed overseas textile manufacturing facility. It was found that the simulation was able to achieve this because of its ability to both store attribute values and to show queuing levels at an individual product level. It was also found that the process of undertaking the simulation project initiated useful discussions regarding the operation of the facility. Discrete event simulation is shown to be much more than an exercise in quantitative analysis of results and an important task of the simulation project manager is to initiate a debate among decision makers regarding the assumptions of how the system operates.

The Case Study

Due to global competitive pressures many garment manufacturers have scaled down or closed their operations in the UK and moved overseas. Due to lower labour costs, in what is a labour intensive industry, production has increased in such areas as Asia. This study involves the design of a proposed overseas textile production facility which supplies garment manufacturers with rolls of material suitable for clothing manufacture. The case study organisation is based in Leicester, UK and produces a range of cotton and lycra textile mixes which are used for garments such as t-shirts and women's tights. In response to the relocation of garment manufacturers overseas and the need to reduce transportation costs of a bulky product the organisation has decided to supplement its UK operations and locate a textile production facility in Sri Lanka. The move will also lower costs due to lower labour rates and permit the design of a more efficient layout in a purpose built factory, as opposed to the current facilities which are placed across a number of locations and buildings within the UK. Previous work involving simulation in the context of textile manufacture includes Gravel and Price (1991), Daly et al. (1995) and Falcone and De Felice (1999). The main

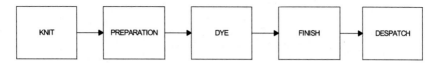

Figure 6.9. Main Stages in Textile Plant Production Process

stages in the production process are shown in Figure 6.9. Each stage is staffed by a locally managed team and all material passes through the knit, preparation, dye, finish and despatch processes.

Each process is now described in more detail

Knit
The knit process takes yarn from the warehouse and knits into 25kg rolls of cloth. A product mix is created by allocating a number of knit machines to a product type in proportion to the mix percentage. Table 6.4 gives a product mix of 7 main product types of different material and weight for the plant configuration to be modelled.

Table 6.4. Product Mix

Product Type	Product Mix %
1	6
2	18
3	18
4	6
5	18
6	28
7	6

After knitting, the 25kg roll is placed in a doff box (rectangular container) and quality examined at one of three knit examination tables. After the quality examination the output from each knit machine is grouped separately to form a batch. This ensures that material within a batch is consistent as it is from the same knit machine and source yarn. With the current configuration this requires 48 batch areas (one for each knit machine). The batch quantity is determined by the dye machine type that this product type has been allocated.

Preparation
The yarn has now been knitted into sheets of material in 25kg rolls. The preparation stage sews these individual rolls together into a batch termed a lane for dying. At this stage certain material types are set to shape using a heated conveyor termed a stenter. The process routes for the seven product types are shown in Figure 6.10.

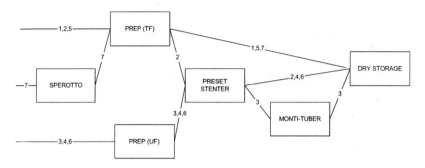

Figure 6.10. Preparation Stage Process Routing

Product types 1 and 5 are processed through the Prep (TF) flapper that sews the individual 25kg rolls from the knit process together. Product 2 requires setting on the preset stenter after sewing. Product 3 is sewn on the Prep (UF), preset on the stenter and then its edges are sewn into a tube on the monti-tuber machine. Product types 4 and 6 are sewn on the Prep (UF) and preset on the stenter. Product 7 is reversed on the sperotto before sewing on the Prep (TF). Material awaiting the sperotto, prep (TF) and prep (UF) processes are stored in the preparation storage area. Material awaiting the preset stenter and monti-tuber processes is stored in the greige (i.e. undyed cloth) storage area. After completion of the preparation process batches are stored in the dry storage area awaiting the dye process.

Dye
The cloth is dyed in one of three shades (dark, medium, light) in a dye machine which resembles a large domestic washing machine. There are four types of dye machine (1 lane, 2 lane, 3 lane and TRD) which have different load capacities. A lane relates to the capacity of one drum within the dye machine. The batch size is derived by multiplying the lane capacity with the number of lanes for each machine. The batch and lane size for each dye machine is given in Table 6.5. For example material allocated for a 3 lane dye machine will consist of 3 lengths (lanes) of 9x25kg rolls stitched together to form a batch. The lanes are transported between production stages as a batch to ensure material from the same source is kept together.

Table 6.5. Batch Size by Dye Machine

Dye Machine Type	Lane Size	Batch Size	Weight (kg)
1 lane	9	9	225
2 lane	9	18	450
3 lane	9	27	675
TRD (2 lane)	8	16	400

The allocation of batches to a dye machine type is undertaken before the knit process and is determined by product type (see Table 6.6). For example batches of product type 1 will only be allocated to the 1, 2 or 3 lane dye machine types. The specific mix of machine types allocated to each batch from this selection is determined by the production schedule to ensure that a balanced load is achieved through all dye machine types. The reason this allocation is determined by the schedule before the knit stage is that batches must contain material from the same knit machine and so cannot be split or combined and the batch size for each dye machine is different.

Table 6.6. Dye Machine Allocation by Product Type

Product Type	Dye Machine Type Allocation
1	1,2,3 lane
2	1,2,3 lane
3	1,2,3 lane
4	1,2,3 lane
5	80% 1,2,3 lane / 20% TRD
6	50% 1,2,3 lane / 50% TRD
7	TRD

The batches arrive at the dry storage area ready for dying by one of the dye machine types. Each batch must now be allocated to particular dye machine of the correct type. For example there may be five 2 lane dye machines available. Dye machines must undertake a lengthy setup process when changing from one dye shade to a different dye shade and so rules have been developed for allocating a batch to a particular dye machine which aims to minimise the number of setups undertaken. Thus each batch must be allocated to the correct machine type for that product type and if possible a machine which has previously been used for the same dye shade.

The rules for allocating a batch in the dry storage area to a dye machine are as follows:

1. If a dye machine of the allocated dye machine type is available, process the batch on this machine. (idle machine)

otherwise

2. If the shade of the last batch on any dye machine of the allocated dye machine type matches the shade of the batch, wait in dry storage until this machine is available and process the batch on this machine. (match shade)

otherwise

3. Wait in dry storage for the dye machine of the allocated dye machine type with the smallest queue. (shortest queue)

The dye process time is dependent on the dye shade and is adjusted for batch weight. Note a 3-lane (675kg) batch on a 3-lane dye machine has a process time equal to a 1 lane (225kg) batch on a 1-lane dye machine. After the dye process the wet material is unloaded into wheeled tubs (1 lane to a tub) and stored in the 'tubs and wet storage' area awaiting the finishing process.

Finishing

The finishing process dries and if necessary sets the shape of the material. The process map for finishing for the seven product types is shown in Figure 6.11.

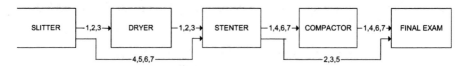

Figure 6.11. Process Route for Finishing

Note that the material is now held in a tub and consists of a lane (eight or nine 25kg rolls) stitched together in a tube. The slitters slit open the tube to form a material length that is flapped to aid drying. Products 1, 2, 3 are passed through a further drying process. All products pass through the stenter machine that sets the material in its final width and weight and provides any chemical finishes that are required. Products 1,4,6,7 are shaped by the compactor. All products are visually inspected individually as lanes and then batched and moved to the finished goods warehouse.

Despatch

Batches of material are held in the finished goods warehouse ready for despatch. Material is held for local, air or sea despatch in the proportion of 50%, 25% and 25% respectively. Local despatches are undertaken daily, Air despatches wait for 5 batches in storage, and then send all in storage 24 hours later. Sea despatches wait for 10 batches in storage, and then send all in storage 7 days later. There may be more than 5 or 10 batches sent by air or sea at a time if further batches arrive in the warehouse after the trigger levels have been reached and before actual despatch.

The Simulation Study

The main stages of the simulation study, based on Greasley (2004), are outlined below.

Study Objectives

The aim of the simulation study is to assist the layout planning activity by estimating the quantity of work-in-progress inventory within the proposed facility to be situated in Sri Lanka. The estimation of inventory levels is critical because the relative bulk of inventory means the amount of floor-space required could be considerable. The need to sink drainage channels for effluent from the knit and dye machines and the size and weight of the machinery involved, means that it would be expensive and time-consuming to change the factory layout after construction.

Data Collection and Process Mapping
The study collected data on the main stages within the production process and modelled seven product types through the factory using data from current experience of operations in the UK and from machine vendor estimates. Each product type has an individual process route and process times which are held on a spreadsheet to allow easy access by personnel. The plant runs on a continuous (7 days a week, 24 hours a day) production cycle with an efficiency rating used to compensate for lost resource availability due to machine setup and other downtime factors.

Building the Model
The simulation model was built using the ARENA (Kelton et al., 2007) Visual Interactive Modelling (VIM) system. Figure 6.12 is a partial zoomed view of the animated model showing the proposed dye production facility. Product batches are represented as coloured circles, with a colour representing each of the seven product types. A letter L, M or D within the circles represents light, medium and dark dye shade respectively. The animation display shows the batches waiting for each dye machine type in the dry storage area and the batches currently being processed on the dye machines.

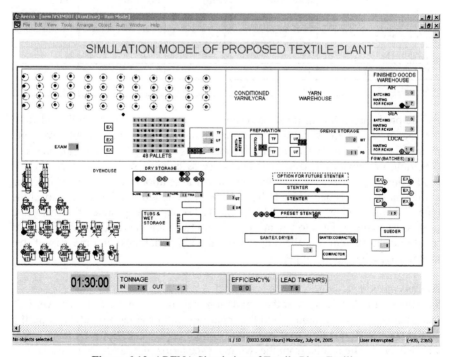

Figure 6.12. ARENA Simulation of Textile Plant Facility

Validation and Verification

Model validation was undertaken by comparing output levels and lead-time measurements with projected figures from the scenarios defined. Detailed walkthroughs of the model were then undertaken with production personnel to validate the behaviour of each production stage in the simulation.

Experimentation and Analysis

The main objective of the simulation study was to predict the amount of work-in-progress (WIP) in the proposed layout under two scenarios of dye capacity which has been identified as the bottleneck constraining overall plant capacity. The experimentation assessed the performance of the production system for two scenarios of a 28 'lane' dye facility and a 39 'lane' dye facility, representing target output capacity levels of 60 tons per week and 100 tons per week respectively. The production system was balanced by restricting the amount of work into the initial knit process to match the capacity of the bottleneck dye facility. The simulation results report on the WIP amount and thus provide a measure of storage area required for all major WIP areas. The dye capacity details for each scenario are given in Table 6.7.

Table 6.7. Dye Capacity for Simulation Scenarios

DYE MACHINE	28 lane scenario		39 lane scenario	
	machines	lanes	machines	lanes
1 LANE DYE	3	3	2	2
2 LANE DYE	5	10	6	12
3 LANE DYE	3	9	7	21
TRD (2 LANE)	3	6	2	4
TOTAL	14	28	17	39

The production system must be balanced before results can be collected from the model. A balanced system will match output level with input level and maintain a steady level of WIP. The dye machines are the production bottleneck or constraint on capacity and so it is necessary to balance the system through this production stage. Because each batch of material is allocated a dye machine type at the knit stage it is necessary to balance the workflow through each dye machine type (i.e. the 1-lane, 2-lane, 3-lane and TRD) separately rather than the dye process as a whole. Thus to balance the system two factors are adjusted, the rate of input into the system (i.e. the output from the knit machines) and the mix of dye machine type by product type.

Implementation

To achieve a balanced system the simulation was run a number of times over a 5-week period (50,400 minutes) and the input rate and product mix adjusted until a steady state WIP quantity (not necessarily equal each week, but on a level trend) was achieved. The results from the two main scenarios of a '28 lane' scenario with 28 dye lane units available and a '39 lane' scenario with 39 dye lane units available, are now described. To calculate the maximum floor-space required for each storage area the maximum WIP amount at that location over 10 simulation replications is recorded. The highest of these 10 measurements is used in conjunction with the floor-space area for the relevant storage type to derive a maximum floor-space area required for each storage location (Table 6.8).

Table 6.8. Floor-space required by storage area

Storage Area	Maximum WIP	Maximum Floor-space (m^2)	
		28 lane scenario	39 lane scenario
Knit Exam	21 rolls	21	21
Knit Batchup	48 batches	120	120
Preparation	37 lanes	87.5	92.5
Greige	41 lanes	65	102.5
Dry Storage	130 lanes	322.5	325
Tubs and Wet Storage	16 lanes	24	32
Stenter	34 lanes	55	85
Compactor	12 lanes	20	30
Final Exam	23 lanes	45	57.5
Finished Goods	40 batches	72.5	100

In summary the simulation was able to estimate the amount of factory work-in-progress (WIP), and thus floor-space area required, under two scenarios of a '28 lane' and a '39 lane' dye configuration. These related to two investment proposals providing two capacity capabilities of 60 tons and 100 tons respectively. The results were obtained by running the simulation over 10 replications and recording the maximum WIP amount at each storage area. The dye stage was identified as the system bottleneck and so in order to balance the system enough work must be provided to the dye facility to ensure the machines are fully utilised, but not too fast a rate that queue build-up occurs. In order to achieve a balanced system the overall rate of work entering the dye facility was controlled by adjusting the rate of output from the knit process. However each product type is pre-allocated a dye machine type at the knit stage which means that the volume through each dye machine type (i.e. 1 lane, 2 lane, 3 lane, TRD) must be balanced separately. This was achieved by adjusting the volume mix on the four dye machine types. Because of this relatively complex procedure a number of simulation runs were necessary to adjust both overall product volume and the mix of products allocated to each dye machine type in order to balance the system. This process is complex because of the lag between the decision taken at the knit stage and the effect of that decision when the batch reaches the dye area. Between these stages each product batch takes an independent route though the intermediate preparation process.

Discussion

In the case study described a simple spreadsheet model could have been used to calculate maximum loading and thus inventory levels in a standard line facility layout. However in this case rules regarding the need to match each batch of cloth with a dye machine type and previous batch shade at the dye storage area, made calculation of possible inventory levels difficult.

A discrete event simulation has the ability to carry information about each entity (by setting an attribute of the entity to a value) which was needed in this case because the allocation of a dye machine type for each batch of material is made before the knit stage of the process in order to ensure consistency of material. Thus when a batch arrives at the dye stage it is allocated a machine based on an attribute set at the beginning of the process. This allocation in turn will determine maximum inventory levels. Also the ability to show queuing behaviour is essential in this instance because of the overall objective of the study is to show the maximum inventory level, and thus floor-space requirement, necessary for the proposed facility layout. Thus discrete-event simulation was chosen because of its ability to both store attribute values and to show queuing levels at an individual product level.

However what also became clear from the case study was that discrete event simulation provided more than a quantitative analysis of floorspace, but generated qualitative data for decision making. Firstly the simulation demonstrated the relationship between decisions made at the knit stage and the effect of these decisions on the downstream dye stage. This issue was important because each production stage (e.g. knit, preparation, dye) had a separate manager responsible for their area of operations. Thus the simulation study underlined the importance of communication and collaboration between these areas in establishing a balanced production cycle and it was proposed to use the visual representation provided by the model as a training tool for the production stage managers in understanding the dynamics and relationships between operations. Secondly at an operational level the need to codify decisions made by personnel in the production process caused their assumptions to be questioned. Specifically as a part of the model data collection process the rules regarding allocation of work to dye machines was classified and formalised after discussion with personnel.

In retrospect it was clear that the study was undertaken on the basis of assumptions about how the process worked that were incorrect. The consequence of this was that additional time was required to form an understanding of the process requiring the project completion date to be extended. These qualitative outcomes are generated because simulation modelling is not just about analysing results from a model, but is a process in which the model may not even be built. What the process of designing and building a model did offer was a way of initiating discussion amongst decision makers about the system in question through such actions as data collection, process mapping and visual inspection of the simulation animation display. Indeed it is not a requirement of a simulation modelling exercise that a model is actually built, but qualitative outcomes from the process mapping stage, for example, could generate useful knowledge. This elicitation of knowledge through the process of conducting a simulation study rather than simply an observation of model results is termed 'simulation for facilitation' by Robinson (2002).

The case study also demonstrated that one of the important factors in achieving confidence in the model from the decision makers involved was the intuitive way in which the discrete event method represents elements such as machines, people and products as recognisable objects. The ability to observe the animation display, which incorporated the CAD drawings of the proposed facility layout, was seen as an essential check of accuracy from the client's perspective.

Summary

In Case Study 4 a model was constructed of a proposed train maintenance depot for an underground transportation facility in the UK. The company who were bidding to operate the depot had traditionally been involved in manufacture and so had no experience of either operating such a facility or meeting the type of performance indicators specified in the service-level agreement. The simulation proved successful in providing a greater understanding of the operation of the depot and the effect of various strategies for meeting demand. The qualitative output of the simulation, in terms of the animation display, proved to be an excellent communication tool during the bid process in helping to show capability to the client in meeting proposed service level targets over time and thus prove service reliability.

In Case Study 5 the simulation achieved its objective in terms of its technical aims, providing an assessment on the performance of the system, and therefore achieving the specific aim of the project. It allowed management to predict if line balancing strategies such as setup reduction and parts sequencing would be sufficient, or if more fundamental changes such as the addition of lines or the replacement of machines was required. However the simulation model failed to realise its potential within a wider organisational basis. That is simulation was used and understood by a small number of management for a 'one-off' project rather than a large cross-section of employees on an ongoing basis. A recommendation was made that an assessment of both quantitative and qualitative outcomes of the use of simulation should be considered when incorporating simulation into change programmes within the organisation. For example qualitative factors such as the potential of using the model building process to gain understanding of a system and the use of animation to communicate ideas should be considered.

In Case Study 6 a discrete event simulation model was used to estimate the size of storage areas required for a proposed overseas textile manufacturing facility. Discrete event simulation was also found to have the ability to facilitate knowledge through the day-to-day process of undertaking the study (for example collecting the data, mapping the processes) and, providing qualitative outcomes (for example an animation of the system incorporating individual elements such as people and materials). In the case study described it was found that the process of undertaking the simulation project initiated useful discussions regarding the operation of the facility covering areas such as the management of the departments and their interrelationships, the accuracy of data held on machine capacity, working practices such as shift patterns and examination of production rules that had evolved over time without any formal assessment of their appropriateness. What the case study does show is that discrete-event simulation in its 'facilitation' role can provide qualitative understanding of behaviour over and above the normal benefits associated with this technique.

References

Bentley, J.P. (1999), *Introduction to Reliability and Quality Engineering*, 2nd Edition, Addison Wesley: Harlow.

Bessant, J.; Gilbert, J.; Harding, R.; Webb, S. (1994), Rediscovering continuous improvement, *Technovation*, 14, 17–29.

Daly, J.S.; Cassidy, B.D.; Kulasiri, G.D. (1995), A simulation model of woollen system carpet yarn manufacture for production planning applications, *Computers and Electronics in Agriculture*, 249–260.

Diamond R., Harrell C.R., Henrikson J.O., Nordgren W.B., Pegden C.D., Rohrer M.W., Waller A.P., Law A.M. (2002), The current and future status of simulation software (panel), *In Proceedings of the 2002 Winter Simulation Conference*, SCS, 1633–1640.

Evans, J.R.; Anderson, D.R., Sweeny, D.J.; Williams, T.A. (1990), *Applied Production and Operations Management*, 5th Edition, West Publishing.

Falcone, D. and De Felice, F. (1999), Dimensioning of a transfer system for the optimization of the flows of the materials in a textile enterprise by means of simulation technique, *Proceedings of the 13th European Simulation Multiconference*, Warsaw, 2, 526–529.

Fitzsimmons, J.A. and Fitzsimmons, M.J. (1994), *Service Management for Competitive Advantage*, McGraw-Hill.

Gravel, M. and Price, W.L. (1991), Visual Interactive Simulation shows how to use the Kanban method in small business, *Interfaces*, 21: 5, 22–33.

Hill, T.J. (1983), *Production/Operations Management*, Prentice Hall.

Hollocks, B.W. (1995), Simulation may be Dangerous – Experimentation Practice and the Implications for Simulation software, *In Proceedings of the 1995 EUROSIM Conference* (Vienna, Sep. 11–15), Elsevier.

Hope, C. and Mühlemann, A. (1997), *Service Operations Management: Strategy, design and delivery*, Prentice Hall.

Imai, M. (1986), *Kaizen: The Key to Japan's Competitive Success*, Random House.

Kachitvichyanukul, V., Henriksen, J.O., Pegden, C.D., Ingalls, R.G., Schmeiser, B.W. (2001), Simulation Environment for the New Millennium (Panel), *In Proceedings of the 2001 Winter Simulation Conference*, SCS, 541–547.

Kelton, W.D., Sadowski, R.P., Sturrock, D.T. (2007), *Simulation with Arena*, 4th Edition, McGraw-Hill, Singapore.

Law, A.M. and Kelton, W.D. (2000), *Simulation Modeling and Analysis*, Third Edition, McGraw-Hill, Singapore.

Law, A.M. and McComas, M.G. (1989), Pitfalls to avoid in the simulation of manufacturing systems, *Industrial Engineering*, May.

Lovelock, C.H. (1988), *Managing Services: Marketing, Operations, and Human Resources*, Prentice-Hall.

McHaney, R. (1991), *Computer Simulation: A Practical Perspective*, Academic Press.

Monden, Y. (1983), *Toyota Production System*, Industrial Engineering and Management Press.

Paul, R.J. (1991), Recent Developments in Simulation Modelling, *Journal of the Operational Research Society*, 42: 3.

Pegden, C.D.; Shannon, R.E.; Sadowski, R.P. (1995), *Introduction to Simulation Using SIMAN*, Second Edition, McGraw-Hill.

Pidd, M. (1998), *Computer Simulation in Management Science*, Fourth Edition, Wiley, Chicester.

Robinson, S. (1994) *Successful Simulation: A Practical Approach to Simulation Projects*, McGraw-Hill, Berkshire.

Robinson S. (2002), Modes of simulation practice: approaches to business and military simulation, *Simulation Modelling Practice and Theory*, 10, 513–523.

Rowe, A.J.; Boulgarides, J.D. (1992), *Managerial Decision Making: A Guide to Successful Business Decisions*, Macmillan.

Schmenner, R.W. (1993), *Production/Operations Management*, Fifth Edition, Macmillan.

Schniederjans, M.J. (1993), *Just-in-Time Management*, Allyn and Bacon.

Seppanen, M.S. (2000), Developing Industrial Strength Simulation Models using Visual Basic for Applications (VBA), *In Proceedings of the 2000 Winter Simulation Conference*, SCS, 77–82.

Shin, D. and Min, H. (1991), Flexible Line Balancing Practices in a Just-In-Time Environment, *Production and Inventory Management Journal*, Fourth Quarter.

Sprague, R.H. (1993), *Decision Support Systems: Putting Theory into Practice*, Third Edition, Prentice Hall.

Stevenson, W.J. (1993), *Production/Operations Management*, Fourth Edition, IRWIN.

Thomas, E. and Dessouky, Y. (2003), The integrated use of process mapping & simulation for a product design process *In Proceedings of the 31st International Conference on Computers and Industrial Engineering*, 251–258.

7

Enabling Simulation – Simulation and OR Techniques

Introduction

This section investigates how simulation can be combined with the operational research techniques of Activity Based Costing (ABC), System Dynamics and Data Envelopment Analysis (DEA). Case Study 7 covers the use of simulation with the technique of ABC. Turney (1996) outlines an ABC model which has two main views (Figure 7.1).

The cost assignment view of ABC allocates costs to activities by identifying *resource drivers* which determine the cost of resources and *activity drivers* which

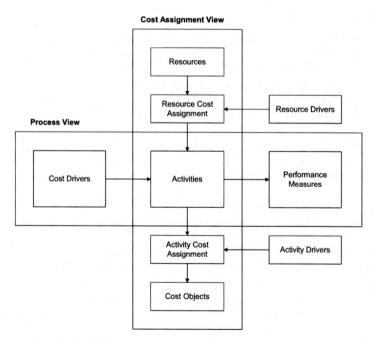

Figure 7.1. The two perspectives of Activity Based Costing

determine the use of these resources. The process view of ABC provides information about the effort needed for the activity termed the *cost driver* and provides performance measures of the outcome of the activities. The cost assignment view can be used to reduce the activity cost by either re-configuring the resources needed for an activity (resource driver) or reducing the amount of resource required (activity driver). The process view provides a link between the inputs needed to perform an activity (cost driver) and the outputs required by the internal or external customer of that activity (performance measure). Thus an investigation of a combination of resource drivers, activity drivers and cost drivers for an activity can improve process performance by identifying why cost has been incurred. A weakness of the ABC approach is that it does not incorporate the effect of the variation of resource usage over time in response to variations in demand. This is particularly relevant in service operations were the intangible nature of the service does not permit storage of a physical product as a buffer against demand fluctuations. Freeman (1992) presents a simulation approach that provides aggregate costs for police activities based on estimates of the average time allocated to these activities. Case study 6 describes a discrete-event simulation implementation of the ABC model. This allows the incorporation of statistical variation in activity duration and thus activity cost, rather than treating the average cost of an activity as a deterministic quantity. The aim is to provide management with information on the variability of cost outcomes in addition to a prediction of the average cost implications of resource decisions.

Case Study 8 presents a simulation study undertaken in conjunction with the system dynamics (SD) approach. The use of a combination of discrete-event simulation and system dynamics has been suggested to analyse operations systems by Fowler (2003), but few examples exist in the literature. Rus et al. (1999) use a combination of system dynamics for project planning and training and discrete-event simulation for more detailed project tracking and control. Martin and Raffo (2001) use system dynamics to represent a software development project environment and a discrete-event simulation to describe the detailed software development process. In addition work has been published on representing discrete events in system dynamics models (Wolstenholme and Coyle, 1980; Wolstenholme, 1982; Coyle, 1985; Curram et al., 2000) and comparing discrete-event and system dynamics models (Crespo-Márquez et al., 1993; MacDonald, 1996; Ruiz Usano et al., 1996; Sweetser, 1999).

Case Study 9 outlines the use of the Data Envelopment Analysis (DEA) technique in the context of the use of discrete-event simulation (Revilla et al., 2003; Chen, 2002). DEA, developed by Charnes et al., 1978; 1979 uses linear programming to assess the comparative efficiency of homogeneous operating units, called Decision Making Units (DMUs). The DMUs are regarded as responsible for converting multiple inputs into outputs. Example DMUs include shops, hospitals, assembly plants and other manufacturing and service units which can be characterised by input-output relationships. Each DMU is seen as consuming a set of resources (e.g. staff time, equipment) to deliver a set of outcomes (e.g. assembled products, serviced customer). The DEA analysis is able to determine a single "technical efficiency" rating for each DMU. The basic concept is that the efficiency of each member of a set of DMUs, the field, is evaluated against its own performance and that of each of the other members of the field. Those DMU that are the most efficient in the combination of all dimensions form an efficiency frontier, and the other units, those less than efficient, are described by a number that indicates their distance from that frontier (Coelli et al., 1998). Thanassoulis (1995) has used DEA to provide an assessment of Police Forces in the UK. The analysis used input variables of the number of crimes in three main categories and the number of officers employed. The output variable was the number of clear-ups of each category of crime recorded. The analysis was able to identify potentially weak and strong Forces on performance, their efficient peers and the level of clear ups that would render inefficient forces efficient.

Case Study 7
A Simulation Analysis of Arrest Costs

Introduction

The following case study describes the use of a simulation study, in conjunction with the ABC approach, to improve process performance at a UK Police Force. The process under investigation is the arrest process, from actual apprehension of a suspect, to processing through a custody suite, to possible interview and court appearance. This process was chosen because it incorporated a reengineering effort in the custody suite and the model could be used to assist analysis in this area. Police management also requested a planning tool for their major human resource, the Police Constable (PC).

The potential long-term increase in cost of such 'monopoly' services where the potential of substitution of people for technology is limited is discussed in Van Reenen (1999). This has implications for Police Services which are characterised by having a large part of their resources available as labour. The wide range of activities that a people-based organisation undertakes often leads to a situation of management of resources by inputs (i.e. budgets) because of the difficulty of classifying the wide range of outputs that police personnel can perform. The amount of resources (people) deployed are based on historical departmental budgets with a large proportion classified as overhead and fixed with an annual addition for inflation. Departments are then managed by tracking variances in expenditure from budgeted amounts.

However resource allocation decisions in the Police Service could also be informed by moving from costs defined in a general organisational budget to an output-oriented budget (Edmonds, 1994). The Activity Based Costing (ABC) approach (Kaplan and Cooper, 1992) allows the user to distinguish between resource usage and resource expenditure, the difference being unused capacity. Once identified this capacity can either be eliminated, reducing costs, or re-deployed, improving effectiveness. This paper will outline the use of ABC as a framework for identifying how cost is generated and will utilise discrete-event simulation to enable the variability of demand and process duration to be incorporated into the cost analysis.

The first step in the study was to estimate the costs associated with the 12 main arrest types of a Police Force. Up to this point, because of the use of budget accounting the actual cost for each arrest type had not been estimated. The purpose of the next stage of the study was to estimate the effect of a change in demand caused by a proposed change in government legislation on resource usage. This requires a change from a consumption model (i.e. product costing) to a spending model (i.e. resource decisions) (Kaplan and Cooper, 1992) termed 'Process Cost Management' by Greenwood and Reeve (1994) who describe an example of its use in a manufacturing context. Their study relies on deterministic cost relationships built within the activity structure of the cost model. This paper implements a cost management system utilising a discrete-event simulation approach which takes into account the variability of process duration and decision rules over time.

Activity Based Costing – Committed and Flexible Resources

ABC does not assume that all organisational costs are variable but uses the concept of committed and flexible resources (Kaplan and Cooper, 1998).

Resources such as a buildings, equipment and energy are termed committed resources when resource expenditure is independent of the actual amount of usage of the resource in any period. This will also cover employees whose costs are constant, independent of the quantity of work performed by them. In service sector examples this capacity is actually acquired before the actual demands for the service are realised and thus the independence in the short-run supply and expense of these resources has led them to be treated as fixed costs. Flexible resources are those which can be matched in the short-term to meet demand. These can include materials, energy, temporary workers hired on a daily basis and the use of overtime that is authorised as needed.

ABC estimates the cost of both committed and flexible resources used by activities and recognises that almost all organisational costs (other than those of flexible resources) are not in practice variable in response to short-term demand fluctuations. Rather committed costs become variable costs over longer time periods by a 2-step procedure.

1. The demand for resources supplied changes because of shifts in activity levels. (through the demand mix changing for example)

2. The organisation changes the supply of committed resources (up or down) to meet the new level in demand for activities performed by resources.

For committed costs to increase, demand exceeds capacity leading to delays and service levels will drop unless more capacity is added. For committed costs to become variable in a downwards direction the organisation must not only recognise the unused costs, but either redeploy resources elsewhere or reduce the level of resources thus reducing cost. Thus 'measuring and managing unused capacity is the central focus of ABC' (Kaplan and Cooper, 1998).

The Simulation Study

The simulation study was conducted following steps based on Pegden (1995):

- Definition of Study Objectives
- System Boundary Definition
- Conceptual Model Formulation
- Data Preparation
- Model Translation
- Verification and Validation
- Experimentation
- Results

Definition of Study Objectives
The arrest process is shown in Figure 7.2 within the ABC matrix.

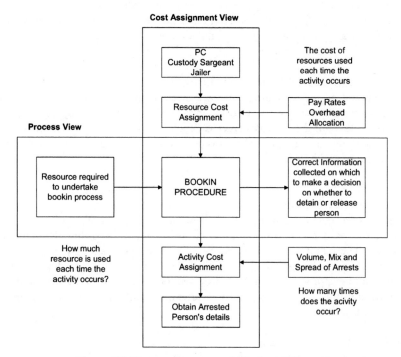

Figure 7.2. The Arrest process within the ABC matrix

From the process view the cost driver is the resource required to undertake each activity. From a cost assignment view the resource driver relates primarily to pay rates for personnel involved in the arrest process. A civilianisation programme (resource driver) and the computerisation of the booking-in process taken within a Business Process Reengineering (BPR) initiative (cost driver) are described in (Greasley and Barlow, 1998). The activity driver relates to the timing and frequency of arrests and provides the focus of this study. The activity driver is dependent on environmental factors such as the crime rate and government policy on crime as well as factors under the control of the Police. This study estimates the personnel cost of each arrest type and then investigates the effect of proposed legislation to extend the opening times of public houses (bars) from 23.00, the current usual closing time in the UK, to midnight. This change represents a change to the activity driver represented in the ABC model.

A second aspect of the study was to consider the custody process from the perspective of each of the three drivers of cost.

- The 'cost driver' relates primarily to the design efficiency of the activities within the custody/arrest process. Costs were computed for each activity (e.g. booking-in, interview) by the simulation by multiplying the activity duration by the appropriate staffing pay rate.

- The 'resource driver' relates primarily to staffing costs for personnel involved in the arrest process. Pay rates (including 'on-costs') were collected for each staff rank (e.g. PC, Jailer) involved in the custody process. Staffing costs were computed by staff rank by the simulation as staff was allocated to activities within the custody process.

- The 'activity driver' relates to the timing and frequency of arrests. Costs were computed by the simulation for each arrest type (e.g. theft, warrant) for the activities engaged by the arrested person within the custody process. The overall level of the activity driver is dependent on environmental factors such as the crime rate and government policy on crime as well as factors under the control of the Police.

System Boundary Definition
The study considers the arrest process, from the initial arrest of a suspect by a PC, through the booking in process at a custody suite to possible detention and interview. The requirement, if necessary, of a court appearance by the PC is also considered. The administration resulting from each arrest is considered both for the PC and the appropriate administration section. The boundary is determined by the need to estimate the main cost of an arrest and particularly the utilisation and cost of PC time.

Conceptual Model Formulation
The main activities in the arrest process are shown in Figure 7.3.

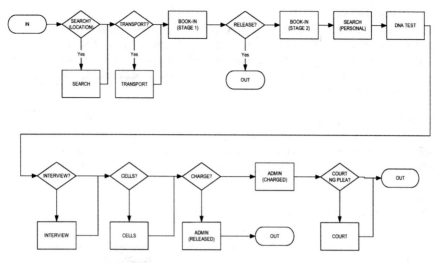

Figure 7.3. Conceptual Model for Arrest Process

The conceptual model of activities is similar to the Process Value Analysis (PVA) method recommended for analysing how activities are related within a process and so improve overall process performance (Ostrenga and Probst, 1992). The first decision point in the arrest process is whether to conduct a search of the location of the arrest. For all decisions during the arrest process a probability distribution is used for each type of arrest at each decision point. The human resource rank required for each process is indicated above the process box. Personnel involved in the arrest process include the PC, custody officer (COF), Jailer and Inspector. The role of each rank is indicated on the conceptual model for each process.

Data Preparation
System Logic The probability of a location search, transportation from arrest location to the custody suite, interview, detention, charge and not guilty plea are estimated for each arrest type. *Demand Pattern* A demand pattern was determined for each of the arrest types for each of the 3 shifts worked at the custody suite. This approach was taken due to the need to estimate resource usage for each of the 3 shifts on duty. It also enabled the change in demand over each day to be modelled more accurately than with a single daily distribution. The user is able to enter the number of arrests (for each arrest type) they estimate will occur in each shift which is translated in the model into an exponential distribution during the shift. The actual arrivals modelled can be compared to the estimates by viewing the simulation display. *Process Timings* A number of distributions were estimated using a curve-fitting program from data collected on activity durations.

Model Translation
The model was built using the ARENA™ modelling system (Pegden et al., 1995), using icons which are connected in a way that resembles the structure of the system. The software permits the user to easily change parameters using custom designed menus. It also incorporates the ability to produce extensive reports in text and graphical format.

Verification and Validation
Verification or 'de-bugging' was carried out making extensive use of the animation facilities to observe the behaviour of components in the system. The model was validated by comparing its performance to the real-world system.

Experimentation Part 1
The first part of the experimentation was undertaken in two stages:

1. Estimate the personnel cost of each arrest type by allocating the pay costs to activities

   As each activity is activated cost is accrued to the appropriate arrest type identified by an attribute of the arrest entity (i.e. arrested person).

2. Estimate the effect of a change in demand pattern (i.e. late drinking scenario) on resource usage.

   In order to estimate changes in resource allocation for the 'late drinking' scenario, estimates of the increase in demand for each relevant arrest type is required. The following increases in arrest occurrences in the appropriate shift (22.00 until 6.00) were estimated by police personnel. Damage 5%, Public Order 7%, Violence 7%, Traffic 20% and Sex 25%.

Results Part 1
To estimate the arrest cost for each arrest type the simulation was run for a period of 50 days. The results are shown in Table 7.1.

Please note at this time that costs include PC, custody officer and jailer pay costs for activities included in the conceptual model only. There will be other costs such as Inspector reviews, meal costs, and PC costs for escorting suspects to court. Other overhead costs are also not included. The figures in Table 7.1 do however provide an indication of the proportion of costs incurred for different arrest types. Of particular interest is the high proportion (32%) of costs incurred from theft offences and the difference between the proportion of arrests on warrant (20%) with the proportion of the cost of these arrests (11%). Thus the overall cost figure provides an indication of the amount of resource allocation for each arrest type both from its frequency and cost per arrest.

Table 7.1. Result of Arrest Cost Simulation

Arrest Type	No. of Arrests	Cost (£)	Cost per arrest	% of total cost	% of arrests
Breach of Bail	60	3222	53.7	5.64	6.16
Burglary	90	7403	82.3	12.95	9.24
Damage	91	5642	62.0	9.87	9.34
Drugs	44	3558	80.9	6.23	4.52
Fraud	17	788	46.4	1.38	1.75
Public Order	63	3048	48.4	5.33	6.47
Robbery	15	969	64.6	1.70	1.54
Sex	16	1558	97.4	2.73	1.64
Theft	267	18303	68.6	32.03	27.41
Traffic	16	749	46.8	1.31	1.64
Violence	104	6558	63.1	11.48	10.68
Warrant	191	5350	28.0	9.36	19.61
TOTAL	**974**	**57148**			

In order to assess the effect on resources of the 'late drinking' scenario the number of PC hours was estimated from the simulation over a period of 50 days. The simulation was run 5 times and the mean of the PC hours estimated for each shift.

Table 7.2. Mean number of PC hours

	PC Hours	Shift 1 6 – 14	Shift 2 14 – 22	Shift 3 22 – 6
Existing drinking hours	Mean	557.0	859.4	962.0
	Std. Dev.	24.3	43.7	36.3
'Late Drinking' scenario	Mean	628.0	851.0	975.6
	Std. Dev.	23.2	41.8	35.7
t value		23.05*	1.22	2.17*

The critical t value for the significance test at 5% level is $t_{8,\,0.05} = 1.86$ *Significant at 5% level

Table 7.2 shows the effect on the mean number of PC hours required of the policy to extend drinking hours from 23.00 to midnight. A one-sided two-sample t test (Oakshott, 1997) shows there is a statistically significant difference (i.e. we can be 95% sure that the change in mean PC hours is not due to random variation alone) in PC hours for shift 1 and shift 3 only. The results show that although the increase in arrests occurs in shift 3 (22.00 until 6.00) most of the additional resources required in response to the predicted increase in arrests will occur in shift 1 (6.00 until 14.00). This is because of delayed activities such as interviews and administration required for court appearances.

Experimentation Part 2

The main aim of this part of the simulation experimentation was to provide an estimate of the costs incurred in the custody process from the perspective of the three drivers of cost.

Activity Driver

The estimates of arrest costs for each arrest type are shown in Table 7.1 and are discussed earlier.

Resource Driver

Table 7.3 shows the staffing costs for the Police Constable (PC), Jailer and Custody Officer derived from the multiplication of pay rate by total activity duration. Further development of the model will also incorporate Inspector costs for reviews of detention periods for example. By estimating resource costs, in this case staff wages, it was possible to estimate the effects and feasibility of proposals to re-allocate and civilianise staffing duties within the custody process. The legal requirements surrounding the custody process constrained the amount of discretion available to personnel in the area and so facilitated the model building process. More loosely based activities would require the use of further experimentation to judge the effect on performance measures. It was envisaged that tasks within the custody area would evolve over time. Greasley (1998) outlines the effect of civilianisation of certain staffing activities which relates to the reduction of costs from a resource driver perspective.

Table 7.3. Cost per Staffing Grade (Resource Driver)

	Average Cost (£)	% of total cost
COSTPC	45171	79.5%
COSTJAILER	8564	15.1%
COSTCOF	3102	5.5%
	56838	

Cost Driver

Table 7.4 shows the costs assigned to each process element within the custody process. The detention process shows a zero cost as the model does not currently include staffing costs for this item. It is envisaged that the model will be developed to include detention costs such as the costs of delivery of meals or release from cells by the jailer. Data is also required on the expected duration of PC time in court to compute court attendance staffing costs. The cost driver relates to the design efficiency of the process elements within the custody operation. Greasley and Barlow (1998) outlines the effect of using Information Technology to improve the design efficiency of the custody process and thus reduce costs from a cost driver perspective. At a higher level of analysis a programme of centralisation of custody suites to smooth demand and focus resources is being implemented, in conjunction with an increase in the level of facilities (e.g. number of cells) to meet predicted peak demand levels. This requires a model providing aggregated costs of custody suites located within a geographical area.

Table 7.4. Cost by Process (Cost Driver)

Process	Average Cost (£)
SEARCH (LOCATION)	5770
TRANSPORT	4192
BOOKIN STAGE 1	5690
BOOKIN STAGE 2	5036
SEARCH (PERSONAL)	3180
DNA	4351
INTERVIEW	11826
DETENTION	0
ADMIN. IF RELEASED	3409
ADMIN. IF CHARGED	13384
COURT	0
	56837

Discussion

At present the custody suite is considered as an essentially fixed cost with an annual budget and there has been no attempt to correlate demand on the facility with costs.

The first stage in the study estimates the arrest costs for each arrest type. This is a function not only of the number of arrests but the likelihood of an arrest leading to interview, detention and court procedures. A Pareto analysis can identify the few activities causing a large proportion of costs (Greasley, 1999). In this case relatively trivial theft offences (usually involving children shoplifting) are causing a heavy workload. This could lead to policies to decrease this workload through crime prevention activities for example.

This paper also describes the use of the model as a resource planning tool for management. Various scenarios can be modelled and the consequences for resource usage (and thus cost) estimated. Here the effect on the implementation of a law regarding 'late drinking' was assessed. These costs will be difficult to model accurately without the use of a simulation model because of the stochastic nature of the system and the interaction between activities within the arrest process. The model can help in providing organisations such as the police with information on which to provide a case to supplying agencies, such as the government, for an increase in resources. Without this information there can be a tendency to assume service providers can 'absorb' the consequences of policy decisions. This can lead to inefficient allocation of resource and poor service quality if the resources available are insufficient.

The model can also assist police management in planning future allocation of staff. With the wide discretion available in how management utilises its main staffing rank (i.e. the PC) it is important that management is aware of the resource implications of their decisions.

In the second part of the experimentation the simulation study has analysed the performance of the custody process from the perspective of the three drivers of cost.

From the activity driver perspective the simulation can provide information on the source of cost by arrest type. Arrest types, such as theft offences, accounting for a high proportion of costs can be targeted to reduce overall costs.

From the resource driver perspective, in professional service organisations such as the Police, staffing costs account for a high proportion of total costs. In this case staffing costs could be reduced by re-allocating tasks from the custody officer to a jailer and investigating the civilianisation of the jailer role.

From the cost driver perspective the effect of increased process efficiency achieved through the redesign of activities can be estimated using the simulation. In this case the introduction of a computerised custody booking-in system was able to decrease the time and cost of the custody operation.

Whichever perspective is used it should be recognised that a reduction in staffing cost obtained will be offset by an equivalent increase in the cost of unused capacity. For example the reduction in cost of custody officer time obtained through re-allocation of tasks and computerisation of the booking-in process will not lead to an overall spending decrease unless the committed cost of this staffing rank becomes variable. This is achieved by a management decision to either re-deploy or eliminate the unused capacity created.

Case Study 8
Using System Dynamics in a Discrete-event Simulation Study

Introduction

The case study company manufacture a range of aluminium gas cylinders for a global market. The cylinders are used in applications such as beverage machines and fire extinguishers. Generally the demand for cylinders is non-seasonal, with the only predictable demand pattern being the annual seasonal variation of heavy and light (weight of cylinders) mixes, associated with demand for beverage (fizzy drink) cylinders. The size of the cylinders range from a diameter of 102mm to 250mm and cycle-times at the various workstations vary accordingly i.e. the bigger the cylinder the longer the cycle-time. The current method of scheduling production is undertaken using a set of complex sequencing rules, built on experience and workstation performance figures. The quality of the products as well as strong market performance has increased the demand for the company's products. In order to achieve high volumes, both factories suffer from a number of common manufacturing problems, including high levels of Work-in-Progress (WIP) inventory, long production lead-times, moving bottlenecks and poor delivery performance. Principally, cylinders can be grouped into 10 different diameter families each supporting 6 separate markets. The smallest design change to a cylinder can necessitate a renaming of the cylinder type. Consequently, over the years, such changes have seen the number of cylinder designs offered to the customer reach approximately 250. The effect on production and capacity planning has been that it is now an extremely complex exercise when attempting to schedule the best mix of products for manufacture.

The Manufacturing Process

Each cylinder is manufactured from a diameter-specific bar of aluminium, which is cut into a billet, machined and then prepared for extrusion. Once extruded, the cylinder has its neck formed and is then heat-treated to age (strengthen) the material. An order-specific internal thread is then machined into the neck followed by a compulsory pressure test sequence. Cylinders are then painted (if requested) and accessories are then fitted. A final internal inspection is then undertaken by an outside testing authority before despatch. The single piece basic cylinder design is shown in Figure 7.4.

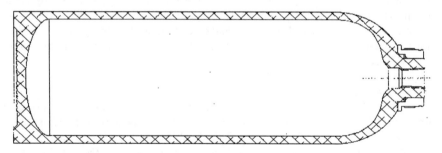

Figure 7.4. Basic Cylinder Design

The main manufacturing stages are now outlined in more detail.

Saw

Specific diameter aluminium logs enter the factory from the metal park. The logs are cut into raw billets of specific lengths through a single saw unit.

Billet Preparation

The next stage of the manufacturing process is the machining of the billets. Billets are given manufacturing codes that describe the size and finish (depending upon cylinder type). Billets are transferred to one of three machine centres depending on size. Generally the larger the diameter of the cylinder, the longer the cycle time. The billet edges are then trimmed to a specific diameter with the ends being skimmed and (if required) tapered. The billets are then stamped for operator identification and then conveyed (in batches) to the etch process.

Etch

A caustic solution is then applied to the whole billet to allow the adhesion of a lubricant to be applied to the pressed face of the billet. The lubricant is then sprayed onto the pressed face and allowed to dry for 24 hours. The billets are then placed into pallets (in batches), ready for extrusion.

Extrusion

The pallets of billets are loaded onto a shelf where the lubricant is cleaned off the edges. The billets are then loaded onto a conveyor with the lubricated face pointing towards the press punch nose. Once the billet is placed into position, pressure builds up within the punch nose ram and the punch nose then forces the billet against a die. Forced into shape by the punch nose, the punch nose reverses its action and the walls of the cylinder are formed. The now hot cylinder is then transported (by a conveyor) for a compulsory wall thickness test and roll stamping (for identification purposes). If successful, the open-ended cylinder is then transported by a conveyor system to the heading dye, where the basic cylinder neck shape is formed. Once the neck is formed, the cylinders are then inverted into heat-treatment baskets. The quantity of cylinders within the baskets depends upon the diameter of the cylinder. Set-up times for each cylinder type to be extruded at the press will depend upon the length, diameter, wall thickness and base design. Each change of design will require a change of tooling. Generally, the cycle time for extruding each cylinder type increases with the overall increase in the size of the cylinder.

Heat Treatment

The baskets are then stacked into oven loading configurations for the heat treatment process which strengthens the cylinders. The capacity of an oven load is inversely related to the size of the cylinder. This has a direct consequence on the overall capacity performance of the heat treatment process. Once the cycle is completed the baskets are unloaded and left to cool (or are quenched in water if required immediately). The basket loads of cylinders are either unloaded manually or through an automatic unloader.

Machining

The cylinders are routed depending upon available capacity and thread finish. A parallel or taper thread is offered into the neck of the cylinder and a collar may also be fitted at the base of the neck for additional strength.

Pressure Test
Once the cylinders have been machined, they are then pressure tested. This is a compulsory test for certification processes. Depending upon capacity, the cylinders have two alternative routes for pressure test. All cylinders are then internally washed to ensure no particles are present. Finally, the cylinders are roll stamped and transported on pallets for inspection or painting.

Paint
Any combination of paint finish can be supplied. Once painted the cylinders are then inspected and made ready for despatch. No cylinders can be released from the factory until they have been inspected by an external and independent inspection body.

The Discrete-event Simulation Study

The use of discrete-event simulation in general as a manufacturing support tool is discussed in Chance et al. (1996). Simulation has been used in production planning applications for a number of years. Galbraith and Standridge (1994) present a case study of the transition of an electronics assembly manufacturing system from a traditional push production control system to a just in time (pull control) system. The use of simulation in moving to a just in time system is also considered by Welgama and Mills (1995). Simulation is used by Spedding and Chan (2001) to analyse the propagation of defectives or errors through a manufacturing system. Dewhurst et al. (2001) discuss extending the use of simulation from analysing and designing manufacturing systems to its use as the basis of a methodology for manufacturing planning. A methodology based on Law and Kelton (2000) was used to construct the model which includes model formulation, model translation, verification and validation, experimentation and reporting of results.

Model Formulation
Figure 7.5 shows the process map that provides the framework for the simulation model. The diagram indicates the flow of material through the system and the decision points for logic flow within the system. For instance all cylinders pass through the saw process stage and then pass through the appropriate billet process dependent on cylinder size. Cylinders with a diameter of 111mm pass through billet 1, cylinders with a diameter greater than 204mm pass through billet 2. All other cylinders pass through billet 3. All cylinders then pass through the remaining stages.

The focus of the simulation study itself was on determining a suitable cylinder sequence in order to optimise system performance. Because this is the main experimental factor the simulation model was formulated in order that a cylinder sequence pattern can be loaded into the simulation model from an EXCEL™ spreadsheet.

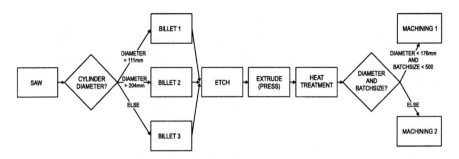

Figure 7.5. Gas Cylinder Manufacturing Process Map

Model Translation

The simulation model was built using the ARENA™ visual interactive modelling (VIM) system (Kelton et al., 2002). This is widely used software in manufacturing and services sectors (for example Greasley and Barlow (1998)) which enables a model to be constructed by placing icons on the screen and sets parameters, such as process times, using dialog boxes. The simulation reads the cylinder data from a spreadsheet file. Various details such as cylinder size and machining production rates are saved as attributes and carried with the cylinder batch as it passes through the simulated production system. Each machine centre is a custom designed program block, built using the 'Professional' version of the ARENA system. This allows the incorporation of a menu system within each work station icon, allowing for easy change of parameters such as machining rates and machine availability.

Verification and Validation

Verification or code debugging was undertaken by extensive use of the simulation animation facilities to observe the behaviour of components within the system. The model was validated by observing performance measures for a cylinder sequence taken directly from the capacity plan. As in all simulation studies the level of detail is dependent on the study objectives. In this case the aim was not to obtain lead time measurements and thus it was only necessary to model up to the machining stage.

Experimentation

The main objective of the simulation study was to investigate the use of production sequencing scenarios in the production planning process. The current method is to base the production sequence on the customer date and order quantity information held within the capacity plan. However this had led to high levels of work-in-progress, long production lead times and thus poor delivery performance. A change suggested by the company was to group together customer orders for either small or large cylinders (each customer tended to only order either small or large cylinder types) in order to reduce machine setup times when moving between cylinder diameters. Two versions of this approach were tested using the model. The 'Small/Large' sequence using a large group size and the

'Small/Large/Small/Large' sequence using a smaller group size. The final scenario tested was to divide the cylinder orders into packets of equal work content, with work content defined as the cylinder cycle time multiplied by the batch size at the extrusion press. The work content of each batch of cylinders was made equal by calculating the average work content for the order sequence and making each batch equal to this. The work was then 'pulled' onto the production system using the extrusion press as the control point. The extrusion press was chosen as the control point as it was identified as the bottleneck process and should therefore set the pace of production (Skrikanth and Cavallaro, 1993). The aim of this approach is to protect the bottleneck from unexpected problems by feeding it with sufficient work and an additional buffer amount.

Results
To investigate the effect of cylinder sequence on the cylinder production cycle time the simulation was run for an extended period until the model reached steady state behaviour. Table 7.5 shows the results for the current sequence, the two suggested sequences of alternating between small and large cylinders and the equal work content sequence described in the experimentation section.

Table 7.5. Sequence Scenario Results

Sequence Scenario	Average Cycle Time (minutes)
Current Sequence	16,256
Small/Large	15,852
Small/Large/Small/Large	16,328
Equal Work Content	15,415

It can be seen that the small/large sequence pattern performed better then the current sequence based on the capacity plan or the Small/Large/Small/Large mix. The improvement in cycle time is small (approx. 2.5%) but would represent a significant increase in cylinder output over a year's production. However using a sequence based on the 'equal work content' principle improves system performance further, leading to a 5% improvement over the current sequence. The approach attempts to provide a smooth workflow to the press by creating batches of equal work content. However a drawback to this approach is the need to calculate the average work content figure and then derive a sequence of cylinder batches that match this figure. In this study a spreadsheet was used to prepare the cylinder sequence. One further difficulty in this approach is the time lag between submitting new material to the production process at the saw process and it reaching the extrusion press. In order to minimise this time lag the control mechanism for feeding material into the system (i.e. the point at which the loading on the extrusion press was monitored) was placed at the etch, rather than the extrusion press itself, to eliminate the 1 day control delay which would occur while material is processed at the etch. The control mechanism could not be placed at the saw or billet process as different sizes of cylinder are processed on different work stations and so the timing of material reaching the press cannot be accurately determined. The simulation study confirmed that the equivalent of 1 day of work-in-progress (WIP) inventory could be eliminated from the production system by moving the control mechanism in this way.

The System Dynamics Study

Although the simulation was able to investigate the operational issues of the manufacturing process itself, a different approach was taken to tackle the wider problem of production planning disruption due to the order scheduling process. This would normally be treated as the 'environment' around which the discrete-event simulation is based. System dynamics (Forrester, 1961) (termed the fifth discipline by Senge (1990)) is an approach for seeing the structures that underlie complex situations and thus for identifying what causes patterns of behaviour. In an organisational setting it is postulated that there are four levels of the systems view operating simultaneously of events, patterns of behaviour, underlying structures and mental models (Maani and Cavana, 2000). Events are reports that only touch the surface of what has happened and offer just a snapshot of the situation. Patterns of behaviour look at how behaviour has changed over time. Underlying structures describe the interplay of the different factors that bring about the outcomes that we observe and mental models represent the beliefs, values and assumptions held by individuals and organisations that underlie the reasons for doings things the way we do them. This framework is now used to analyse the case study scenario.

Events

A recurring event was that of missing customer order dates. This was leading to dissatisfied customers and an increasing reliance on a few 'strategic' customers who were using their buying power to negotiate price reductions and so reduce profitability.

Patterns of Behaviour

Orders are made to customer demand (i.e. not supplied from stock) and are placed on to the capacity plan over a number of months before the actual order is manufactured. The position of an order in the plan is a result of negotiations between the customer, the sales manager who liaises with the customer and a capacity planner. A senior manager is also involved for 'strategic orders' which are deemed to be of particular importance to the company. There is no fixed definition of what makes an order 'strategic'. The production plan for an order is only fixed when the order enters the manufacturing process, approximately 7 weeks before delivery to the customer. Because of the wide variety of products offered, the extended amount of time an order is on the capacity plan and the changing capacity situation lead to a great deal of uncertainty in the capacity planning process. In particular the process of expediting 'strategic' orders has led the company to be labelled by many 'non-strategic' customers as unreliable in terms of delivery performance. This behaviour has increasingly led to some customers over-ordering or ordering in advance to ensure on-time delivery leading to a capacity plan which overstates the actual capacity requirements. Also due to poor delivery performance some customers have chosen to move to alternative suppliers which have led to an increasing proportion of the output being dedicated to 'strategic' customers. This has led to an even greater use of expediting as the 'strategic' customers become an ever increasing proportion of the company's output.

Underlying Structures

One of the tools of system dynamics are system archetypes which are certain dynamics that recur in many different situations. An archetype consists of various combinations of balancing and reinforcing loops. The "fixes that fail" archetype (Kim, 1992) describes a situation in which a solution is quickly implemented that alleviates the symptom of the problem, but the unintended consequences of the "fix" exacerbate the problem. Over time the problem symptom returns to its previous level or becomes worse. An example of the "fixes that fail" archetype is that of 'expediting customer orders' (Figure 7.6).

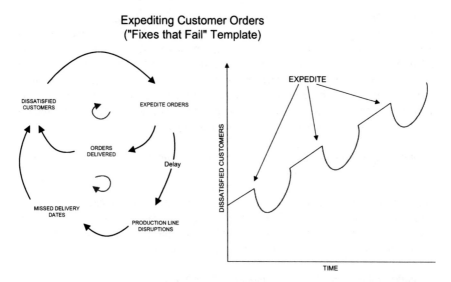

Figure 7.6. Expediting Customer Orders Systems Archetype (Source: Kim (1992))

Figure 7.6 shows how in an effort to ensure on-time delivery an order is expedited resulting in a satisfied customer. However the disruption caused by this policy has led to the need to expedite further orders and a feedback loop is established where more expediting leads to disruptions, leading to missed delivery dates leading to dissatisfied customers and more expediting. The reason this situation occurs is that the pain of not doing something right away is often more real and immediate than the delayed negative effects. The situation is exacerbated by the fact that the reinforcing nature of unintended consequences ensures that tomorrow's problems will multiply faster than today's solutions. In other words solving one problem today, will not create another one tomorrow, but will create multiple problems. Breaking the archetype requires the acknowledgement that the fix is merely alleviating a symptom and making a commitment to solve the real problem now. In this case the idea of 'strategic' orders is creating a self-fulfilling prophecy in which eventually all orders will be 'strategic'. In this situation the company will supply to fewer and fewer customers, who if they realise their market power, can exert substantial pressure on prices and profit margins.

Mental Models

This concerns the most 'hidden' level of system dynamics and concerns assumptions, beliefs, and values that people hold about the system. In order to investigate these issues which influence the accuracy of the demand forecast, a questionnaire was administered to ten sales managers who provide the interface between the customer and the capacity plan. The questionnaire consists of a number of closed questions with space for open-ended comments after the answers. The results of the questionnaire are presented in appendix A and a summary of outcomes is presented below codified into three themes:

Theme 1: Changes to the Order Specification (Questions 1–4)
The sales managers report that customers request production space usually within a 12 month period. 8 sales managers request this production space in advance often. Space is usually requested by medium to large size companies. 7 sales managers have experience of customers often or always changing the size and/or delivery date of the order. Up to 3 months notice is given for this and 4 sales managers report disruption to other orders when they do change the production schedule to accommodate this. As one sales manager explained,

"Some of my customers tend to want their delivery postponing when cylinders are due out in two weeks time. By this time it is too late"

The reason for making these changes is clear in that 8 sales managers report they gain orders from accommodating these changes whilst only 2 claim orders are lost. One reason why customers may do this is that they have a misunderstanding about the production process, according to one sales manager,

"Some customers still believe cylinders to be 'off the shelf'"

Theme 2: Capacity Planning Rules (Questions 5–9)
9 sales managers always consult the capacity planner before quoting a delivery date and 9 sales managers never or rarely are refused space by the capacity planner for the order. Only 6 sales managers understand the current capacity planning rules, and only 5 sales managers actually work to these rules. The

following quotes underline the current misunderstanding of the capacity planning rules,

"Don't know if any hard fast rules exist"

"Rules changed three times since we started forecasting. Sometimes there is a lot of confusion about what is requested"

"Don't know of any rules. The larger customers tend to plan their requirements for the whole year"

Of the 4 sales managers who don't understand the rules all agree it would help them in their task if they did.

Theme 3: Priority Customers (Questions 10–11)
All 10 sales managers have priority customers and 9 sales managers often or always have priority orders scheduled. The need to satisfy 'strategic' customers is underlined by the comment from one sales manager:

"Companies X, Y and Z more or less always get what they want"

In summary from a customer perspective the late changing of orders is accepted as if it was part of the normal "trading" practices with the company. Poor delivery performance has "labelled" the company as unreliable. Consequently, the customers regard themselves as being in a strong bargaining position. Customers prefer to book in advance provisional orders as the lead-time is long. Some customers believe the company makes to stock which in reality has never happened. A feeling of confusion exists over the capacity planning rules. The effect of this is that rules are "made up" to suit the needs of the individual with the consequence that order quantities are over-booked, enquires and reservations are booked as "provisions" for their customers. Non-standard practices are accepted as normal as this enables some element of control over a system that is poorly understood. Finally a priority system operates in order to ensure what the sales managers consider are the most important customers receiving on-time deliveries.

Senge (1990) states that most people assume cause and effect are close in time and space. Thus a fixation on short term events will fail to uncover the longer term pattern of behaviour caused by their actions – in this case the longer term effect on scheduling stability of short-term expediting decisions. In order to break the cycle both the belief and assumptions about the planning process had to be addressed and an even greater understanding of the systemic structures was needed in order to understand the relationship between short-term fixes and a longer-term drop in performance. In order to achieve this, meetings were held with the sales managers and use was made of the archetype diagrams and the results of the questionnaire to discuss the effect of the current planning rules. As a result of these meetings the following policies were agreed:

- Quotes for delivery times to 'strategic customers' would not disrupt the current production plan.
- Delivery times would be maintained and not brought forward if they caused disruption to the current production plan
- Sales to liaise closely with manufacturing to gauge realistic delivery performance.
- Manufacturing to liaise closely with sales regarding up to date production schedule information

Discussion

The case study has described a DES study of a production facility that provided suggestions for improvement in terms of production sequencing. As the project progressed it became clear that although the DES would meet the objectives of the study in a technical sense, the organisational problem of 'delivery performance' would not be solved by the DES study alone. After a request by the author a wider brief to the project was agreed involving an assessment of the order scheduling process undertaken by the sales managers.

A system dynamics approach is used to provide a framework for understanding why things are happening in the way they are by identifying the structure behind behaviour. This differs from the discrete-event simulation approach which generally replicates the structure and identifies behaviour under a number of scenarios. A system dynamics approach involves identifying an archetype which describes the systemic structure of the system. It also assesses the underlying beliefs and assumptions of the participants. In this case it was found that the system dynamics approach provides the following advantages:

- Expands the DES model of the production system to a model that incorporates the role of decision makers, thus providing a deeper understanding of system behaviour and the reason for missed deliveries.
- Treats the cause of the production problem rather than the symptom. This led to recognition of the relationship between missed deliveries and the actions of sales and production personnel.
- Demonstrates, using the system archetype, that a more long-term view of improvements following on from change was required rather than the 'quick fix' solution of expediting orders.
- Shows the need for personnel to consider their own actions rather than blame others for outcomes. All participants in the sales and production process need to consider how their actions affect delivery performance for all customers.

Of interest in the case was how the DES study led to a wider analysis using system dynamics. Initially in the case study the technique of discrete-event simulation was commissioned for a 'hard' technical analysis of a production system. However it has been noted that the DES technique can incorporate qualitative aspects in its analysis (Robinson, 2001; Swenseth et al., 2002). For example the use of process maps and the visual animation display can provide a forum for discussion and understanding. In this case discussion between participants regarding the organisational issues surrounding the production planning system, triggered by the DES study, led to an analysis using the system dynamics technique. This analysis would be unlikely to have occurred without the DES study because the SD technique was unfamiliar to the organisation and initially the problem was seen as a technical issue requiring a DES solution.

Thus it is argued that the qualitative outcomes of DES should be seen as an important part of the DES technique and the system dynamics approach can provide a useful addition to the toolkit of DES practitioners.

However there are limitations to both the DES and SD approaches which may warrant the use of further techniques. For example although DES and SD can make some attempt to incorporate the views of participants they may not be able to grasp the complexity of social reality necessary to ensure a correct analysis is made. Further 'systems approaches' such as Soft Systems Methodology (Checkland, 1981) may be needed in such situations (Jackson, 2003).

Case Study 9
The Use of Data Envelopment Analysis in a Discrete-event Simulation Study

Introduction

This case study is based on Thanassoulis (1995) which assessed police forces in England and Wales. Unlike most European countries, there is no national police force in these countries. Policing services are delivered by 43 autonomous forces which vary in size from around 1000 to 7000 officers. For the purposes of the analysis the London Metropolitan Police Force and the City of London Police Force were not deemed comparable to the other Forces because of their special duties such as the provision of diplomatic protection, so this leaves 41 Forces to be assessed. The analysis took the steps of a preliminary data analysis, an initial DEA assessment and a further investigation to refine the results.

Preliminary Data Analysis

Common to all uses of DEA are the choices of DMUs to be evaluated and the input and outputs from which evaluations are to be effected (Cooper and Tone, 1997). In this case each Force was defined as a DMU and the following input-output variables were defined:

- Number of crimes of each crime category recorded (input)
- People employed at each Force (input)
- Number of clear ups of each crime category recorded (output)

Categories for crimes and clear ups where defined as violent crimes, burglaries and 'other' crimes. The main aim of the preliminary analysis was to identify any associations between the above variables which would have implications for how they are used in the DEA assessment. One aspect of the analysis was an investigation to determine whether there was any strong positive association between the clear up rates of different crime categories. In this case none of the correlation coefficients determined by the analysis were found high enough to warrant dropping any crime category from the DEA assessment.

The Initial DEA Assessment

The details of the DEA computations are shown in Thanassoulis (1995). An efficiency rating was derived to measure the extent to which clear up levels at each Force can improve if they were to perform efficiently relative to other Forces. For example the efficiency rating of 65.03% for Force 'F21' reflects the fact that its clear ups in any one crime category are at best only 65.03% of the level that could be achieved if the Force had been relatively efficient.

Further Investigations

Three aspects of the analysis of the initial results are presented to provide a guide to simulation practitioners on some of the issues that arise in a typical DEA analysis.

1. Genuinely Efficient Resources

At present the results show 13 Forces with an efficiency rating of 100%. As with all DEA analysis there must be a consideration that what has made the DMU 'technically' efficient may constitute a performance that is not very acceptable in practice. Two points were of concern in this particular case. Firstly the Force may be at a part of the efficient boundary characterised by unacceptable marginal rates of substitution between outputs, say valuing one or more outputs excessively while giving negligible value to other outputs. Secondly, the efficient boundary in DEA is established by comparing the observed input-output correspondences at various Forces. If the 'input-output mix' of an efficient force is unusual and not found in many Forces, the Forces position on the efficient boundary will be reflecting its unusual input-output mix rather than its efficient performance relative to other Forces. The details of the investigation into these issues and the steps taken to take account of these factors in the subsequent analysis are detailed in Thanassoulis (1995).

2. Input – Output Weights

The value of clearing up different types of crime was felt to be unequal. Therefore to ensure that efficiency ratings for the Forces were based on output weights that reflected this concern a value system was imposed on the efficiency ratings. The value system specified the weighting on 1 violent crime clear up => 10 x weighting on 1 burglary clear up => 2 x weighting on 1 'other' crime clear up. The main effect of the weight restrictions is to drop the number of Forces that could be deemed to be 100% efficient from 13 to 3, namely Forces 'F30', 'F7' and 'F19'.

3. Efficient Peers

There was some concern that not all inefficient Forces could be directly compared with efficient Forces due to factors such as relative economic status. To overcome these Forces were grouped into families based on socio-economic and settlement pattern indicators so a Force could be compared with other Forces in its own family. Thanssoulis (1995) outlines the use of a set of performance indicators to contrast inefficient Forces with a DEA peer from their own family.

Simulation Case Study: The Redesign of the Crime Arrest Process at a UK Police Force

This section outlines part of a simulation study of a UK Police Force concerning the redesign of the crime arrest process. The custody suite is the point of call for any crime arrests carried out by police officers in the field. The custody operation involves taking details of the arrested person by a designated custody officer and then the possible detention and interview of that person at a later date. The objective of the study was to identify a custody of prisoner process which is legitimate and best practice, and provide recommendations for computerisation. A particular focus of the study was on reengineering the use of human resources within the area.

The personnel involved with the custody of prisoner process are Superintendent, Inspector, Custody Sergeant (Officer), Police Constable and Jailer or trained Civilian. Each of these is ranked from top to bottom – top being superintendent. The premise being that the high rank can always legitimately perform the function of any person beneath them in the order of ranking, however a low rank cannot legitimately perform any function of a person above them. The emphasis of the study was on the allocation of staffing roles to custody operations. The aim was to greatly reduce staffing costs and increase the time available for 'front-line' police work. To achieve maximum efficiency it is necessary to maximise the functional use of the lowest grade of employee, which in this case is a civilian jailer. Thus the study concentrated on the effect of re-allocating tasks from the Custody Officer to the Jailer and then on the feasibility of the proposed 'civilianisation' of the Jailer role.

More details of the construction of the simulation using the Arena simulation system (Kelton et al., 2002) are found in Greasley (2000) and Greasley and Barlow (1998). The simulation was run with the present task allocation of custody officers and jailers within the system. The simulation was then run re-allocating tasks from the custody officers to the jailers. To do this a radical re-think of the custody officer role was required. This role contains important tasks in that the custody officer must ensure that the required information and correct actions are taken in respect of a person brought in by a Police Constable for arrest. The role had evolved over a number of years and it was assumed that most tasks carried out by the custody officer could only be carried out by them or by a person higher in the hierarchy (e.g. a superintendent). However after a study of legal requirements it was found that this was not the case. In fact many of the tasks were of an administrative nature and did not require the authorisation of the custody officer. The custody operation was redesigned from scratch from a process perspective. A procedure was designed which involved the custody officer carrying out a series of tasks required by law and then handing over the 'case' to the jailer to process.

It was found by running a simulation of the new process design that custody officer utilisation has dropped from 29% to 9% while jailer utilisation had risen from 25% to 67%. Police Constable utilisation had fallen from 150% to 125%. Thus the study showed that by re-allocating tasks as proposed provided a significant shift in resource utilisation away from the custody officer and on to the jailer. This re-allocation of tasks down the hierarchy could bring significant savings in staffing costs provided the staff time saved can be utilised effectively. Furthermore the results show that the jailer role can accommodate the re-allocated booking-in tasks, in addition to present duties, assuming the demand and process durations used in the model.

Discussion

The two case studies outlined both concern the crime arrest process in the UK Police Force. The DEA analysis assesses performance at a Police Force unit level and uses the performance measure of the number of crimes cleared up. The simulation analysis assesses the performance of an individual unit in terms of the utilisation of staffing ranks involved in the crime arrest process. These cases are now used to provide an example of the use of simulation and DEA analysis in process redesign (Figure 7.7).

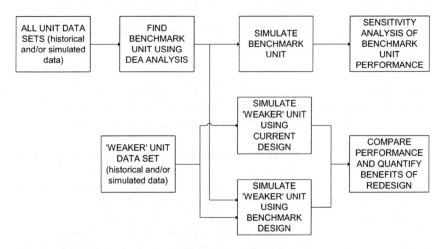

Figure 7.7. A three-stage model of the use of DEA and simulation

In the DEA analysis of the Police Force historical data sets of published information of Police performance were utilised. One aspect of DEA in the use of data from operating units is that an example may not exist or be available for each combination of input variables. Examples of the use of simulation to generate a data set for the DEA analysis include McMullen and Frazier (1998) who use DEA to compare assembly line balancing solutions. The input variables in this case are crew requirement and equipment requirement and the output variables are cycle time ratio and percentage of on-time completions. Because data sets did not exist for each combination of crew and equipment being investigated a simulation model was built which could generate the output measures for a number of input variable scenarios. Another example is Braglia and Petroni (1999) who use DEA to compare dispatching rules in a dynamic hybrid flow shop environment.

What can be seen from the case study DEA analysis is that the DEA procedure is not simply about entering data into a formula and observing the solution? Further analysis is required of the results in order to determine if refinements are necessary to the experimental design in order to provide an intuitive and practicable solution. Although there are many variations to the basic DEA approach to take account of the different contexts in which the tool is used, the flexibility of the approach in incorporating these refinements should be seen as an advantage by the simulation analyst.

What the DEA analysis can provide is an indication of where benchmark process designs are situated across comparable business units. A further use of the simulation model would be to test the robustness of the benchmark unit performance by conducting a sensitivity analysis of the benchmark design using stochastic data.

In this case the DEA analysis could have been used to assist in the redesign process conducted in the simulation study. Although the focus of the simulation study was to increase Police Officer availability it is likely that one of the reasons for the high performance of benchmark forces in clearing up crime is the relatively efficient use of their Officer resource. Thus their crime arrest process design could have been used to inform the simulation redesign study. In general once the DEA analysis has identified the benchmark business unit the ability of simulation to operationalise change can be used as a mechanism to transmit the practices of these units to weaker performing units. This is accomplished by the procedure of process mapping and building a simulation model of the weaker unit to the process design of the benchmark unit. The model can then compare performance of the current and benchmark process designs. Quantifying improvement in this way provides a useful driver to any process change initiative that is required to bring the performance of weaker units up to the best in class.

Summary

Case Study 7 presents a simulation model which provides a suitable platform on which to conduct an ABC analysis due to the ease in which costs can be attached to activities within the model. The ABC approach allows the actual costs to be traced to activities and so enables better resource allocation decisions. In this case the relative cost of each arrest could be clearly seen. Reversing the flow of information allows the user to assess the effect of a change in the activity level on costs. In this case a predicted change in the frequency and timing of arrests due to a proposed change in the 'late drinking' law could be assessed.

Case Study 8 provides an example of the use of simulation in conjunction with the technique of system dynamics. It is shown that the system dynamics approach is particularly appropriate in analysing factors impacting on the organisational context of a simulation study and thus could be used to maximise the benefits of simulation.

Case study 9 shows how the technique of data envelopment analysis (DEA) may be a useful addition to the toolkit of a simulation analyst in that it is able to rank the relative performance of units across multiple input and output measures. Some or all of the data sets needed for the analysis may be generated by the simulation method. The results of the DEA analysis may be used to identify benchmark practices. The simulation can then operationalise the results of the DEA analysis by providing a model of the benchmark unit which can be used to predict performance under a number of scenarios. Furthermore by comparing the performance of a 'weaker' business unit under current and benchmark process designs the benefits of change can be quantified and used as a driver for the redesign effort.

References

Braglia, M. and Petroni, A. (1999), Data envelopment analysis for dispatching rule selection, *Production Planning and Control*, 10, 454–467.

Chance, F., Robinson, J. and Fowler, J. (1996), Supporting Manufacturing with Simulation: Model design, development, and deployment, *Proceedings of the 1996 Winter Simulation Conference*, SCS, 114–121.

Charnes, A., Cooper, W.W., Rhodes, E. (1978), Measuring the efficiency of decision making units, *European Journal of Operational Research*, 2, 429–444.

Charnes, A., Cooper, W.W., Rhodes, E. (1979), Short communication: Measuring the efficiency of decision making units, *European Journal of Operational Research*, 3, 339.

Checkland, P.B. (1981), *Systems Thinking, Systems Practice*, John Wiley & Sons, Chichester, UK.

Chen, T. (2002), A comparison of chance-constrained DEA and stochastic frontier analysis: bank efficiency in Taiwan. *J Opl Res Soc,* 53, 492–500.

Cooper, W.W. and Tone, K. (1997), Measures of inefficiency in data envelopment analysis and stochastic frontier estimation, *European Journal of Operational Research*, 99, 72–88.

Coyle, R.G. (1985), Representing Discrete Events in System Dynamics Models: a theoretical application to modelling coal production, *Journal of the Operational Research Society*, 36, 307–318.

Crespo-Márquez, A., Usano, R.R., and Aznar, R.D. (1993), Continuous and Discrete Simulation in a Production Planning System. A Comparative Study, *International System Dynamics Conference*, Cancun, Mexico.

Curram, S., Coyle, J., and Exelby, D. (2000), Planes, Trains And Automobiles... A System Dynamics Model With Discrete Elements, *18th International Conference of the System Dynamics Society*, Bergen, Norway.

Forrester, J.W. (1961), *Industrial Dynamics*, Productivity Press, Portland, OR.

Fowler, A. (2003), Systems modelling, simulation, and the dynamics of strategy, *Journal of Business Research*, 56, 135–144.

Freeman, J.M. (1992) Planning Police Staffing Levels, *J. Opl Res. Soc.*, 43, 187–194.

Galbraith, L. and Standridge, CR. (1994), Analysis in manufacturing systems simulation: A case study, *SIMULATION*, 63: 6, 368–375.

Greasley, A. and Barlow, S. (1998), Using simulation modelling for BPR: resource allocation in a police custody process, *International Journal of Operations and Production Management*, 18: 9/10, 978–988.

Greasley, A. (1999), *Operations Management in Business*, Stanley Thornes Ltd.

Greasley, A. (2000), A simulation analysis of arrest costs, *J Opl Res Soc*, 51, 162–167.

Greenwood, T.G. and Reeve, J.M. (1994), Process Cost Management, *Cost Management*, Winter, 4–19.

Jackson, M.C. (2003), *Systems Thinking: Creative Holism for Managers*, John Wiley & Sons Ltd, Chichester, UK.

Kaplan, R.S. and Cooper R. (1992), Activity-Based Systems: Measuring the Cost of Resources, *Accounting Horizons*, Sept, 1–13.

Kaplan, R.S. and Cooper, R. (1998), *Cost and Effect: Using Integrated Cost Systems to Drive Profitability and Performance*, Harvard Business School Press, Boston.

Kelton, W.D., Sadowski, R.P. and Sadowski, D.A. (2002), *Simulation with Arena*, 2nd Edition, McGraw-Hill, Singapore.

Kim, D.H. (1992), Systems Archetypes: Diagnosing Systemic Issues and Designing High-Leverage Interventions, *Toolbox Reprint Series: Systems Archetypes*, Pegasus Communications, 3–26.

Law, A.M. and Kelton, W.D. (2000) *Simulation Modeling and Analysis*, Third Edition, McGraw-Hill, Singapore.

Maani, K.E. and Cavana, R.Y. (2000), *Systems Thinking and Modelling: Understanding Change and Complexity*, Pearson Education, New Zealand.

Martin, R. and Raffo, D. (2001), Application of a hybrid process simulation model to a software development project, *The Journal of Systems and Software*, 59, 237–246.

MacDonald, R.H. (1996), Discrete Versus Continuous Formulation: A Case Study Using Coyle's Aircraft Carrier Survivability Model, *International System Dynamics Conference*, Cambridge, Massachusetts.

McMullen, P.R. and Frazier, G.V. (1998), Using Simulation and Data Envelopment Analysis to Compare Assembly Line Balancing Solutions, *Journal of Productivity Analysis*, 11, 149–168.

Oakshott, L. (1997), *Business Modelling and Simulation*, Pitman Publishing.

Ostrenga, M.R. and Probst, F.R. (1992), Process Value Analysis: The Missing Link In Cost Management, *Cost Management*, Fall, 4–13.

Pegden, C.D., Shannon R.E. and Sadowski R.P. (1995), *Introduction to Simulation using Siman*, Second Edition, McGraw-Hill.

Revilla, E., Sarkis, J., Modrego, A. (2003), Evaluating performance of public-private research collaborations: A DEA analysis. *J Opl Res Soc,* 54, 165–174.

Robinson, S. (2001), Soft with a hard centre: discrete-event simulation in facilitation, *Journal of the Operational Research Society*, 52, 905–915.

Ruiz Usano, R.; Torres, J.M.F.; Marquez, A.C. and Castro, R.Z.D. (1996), System Dynamics and Discrete Simulation in a Constant Work-in-Process System: A Comparative Study, *International System Dynamics Conference*, Cambridge, Massachusetts.

Rus, I.; Collofello, J.; Lakey, P. (1999), Software process simulation for reliability management, *The Journal of Systems and Software*, 46, 173–182.

Senge, P.M. (1990) *The Fifth Discipline: The Art and Practice of The Learning Organization*, Random House, London.

Skrikanth, M.L. and Cavallaro, H.E. (1993), *Regaining Competitiveness: Putting the Goal to Work*, Second Revised Edition, North River Press.

Spedding, T.A. and Chan, K.K. (2001), System level improvement using discrete event simulation, *International Journal of Quality and Reliability Management*, 18: 1, 84–103.

Sweetser, A. (1999), A Comparison of System Dynamics (SD) and Discrete Event Simulation (DES), *17th International Conference of the System Dynamics Society and 5th Australian & New Zealand Systems Conference*, Wellington, New Zealand.

Swenseth, S.R., Olson, J.R., Southard, P.B. (2002), Extending product profiling through simulation, *International Journal of Operations and Production Management*, 22: 9, 956–971.

Thanassoulis, E. (1995), Assessing police forces in England and Wales using Data Envelopment Analysis, *European Journal of Operational Research*, 87, 641–657.

Turney, P.B.B. (1996), *Activity Based Costing: The Performance Breakthrough*, Kogan Page.

Welgama, PS. and Mills, RGJ. (1995), Use of simulation in the design of a JIT system, *International Journal of Operations and Production Management*, 15: 9, 245–260.

Wolstenholme, E.F. and Coyle, R.G. (1980), Modelling Discrete Events in System Dynamics Models", *Sixth Intl. Conf. on System Dynamics*, Paris.

Wolstenholme, E.F. (1982), Modelling Discrete Events in System Dynamics Models, *Dynamica*, 6, part 1.

Index

Activity Based Costing (ABC) 119, 121–122
Activity cycle diagrams 41
Agent-based simulation 17
Animation inspection 46
ANOVA 52
ARENA 15, 27

Balanced scorecard 78–79
Batch means analysis 54
Believability 48
Business Process Modelling (BPM) 2
Business Process Reengineering (BPR) 72
Business Process Simulation (BPS) 2, 73

Common Random Numbers (CRN) 52
Communication tool 104
Comparing alternatives 50
Conceptual model 93
Conceptual validation 46
Confidence interval 49
Critical success factors 76–78
Custody of prisoner process 69

Data collection 39
Data Envelopment Analysis (DEA) 146
Decision Support System (DSS) 105
Discrete-event simulation 133
Distribution 44

Empirical distribution 44
ESIA 82
Estimation 43
Experimentation 48
Expertise transfer 104
Extended run length 53

Facility design 106

Historical data points 44
HR division case study 75
Human behaviour 13
Hypothesis testing 50

Implementation 55
Input data 42
Interdependence 3

Line balancing 100

Managerial involvement 55
Marking guide 78–79
Model design 45

Non-terminating systems 52

One-off technique 104
Operational involvement 56
Operational validity 47
Organisational use 97

Paired t-test 51
Performance measurement 83
Police force 77
Presentation of results 54
Process improvement 79
Process mapping 41–42, 63–64, 73, 78

Road traffic accident 63

Scoring system 80, 82
Service reliability 89
SIMAN/CINEMA 15
SIMUL8 27
Simulation
    Discrete-event 41
    Hardware 23

Languages 24
Level of usage 33
Manufacturing applications 12
Project management 34
Project proposal 33, 38
Service applications 13
Software 23, 28
Sponsor 22
Starting conditions 53
Structured walkthrough 45
Surveys 7, 9
System dynamics 15, 138, 141

Terminating systems 49
Test runs 45

Theoretical distribution 43
Trace analysis 46
Training 23, 28

Validation 44, 46
Variability 2
Variance Reduction Techniques (VRT) 52
Verification 45
Visual Interactive Modelling (VIM) 24

Warm-up period 53
WITNESS 27

Languages, 24
Level of tasks, 43
measurements, physical, 12
Risk management
Traca's proposal, 32
risk applications
Software, 25, 95
Sponsor, 25
Starting condition, 53
Structured walled rough, 95
Surveys, A.9
System dynamics, 18, 19, 111

Terminating systems, 44
Test runs, 45

Valuation, 44
Verification, 2
Validity test

Variation, 4
Visual literacy

WBS

Lightning Source UK Ltd.
Milton Keynes UK
UKOW05n1322240214

227031UK00001B/13/P